An introduction to
CENTRIFUGATION

An introduction to
CENTRIFUGATION

T.C. FORD

Department of Biology, University of Essex,
Wivenhoe Park, Colchester CO4 3SQ, UK

and

J.M. GRAHAM

MIC Medical Ltd, Merseyside Innovation Centre,
131 Mount Pleasant, Liverpool L3 5TF, UK

βIOS
SCIENTIFIC
PUBLISHERS

© BIOS Scientific Publishers Limited, 1991

First published in the United Kingdom 1991 by
BIOS Scientific Publishers Limited,
St Thomas House, Becket Street, Oxford OX1 1SJ

A CIP catalogue record for this book is available from the British Library.

ISBN 1 872748 40 6

Printed by Information Press Ltd, Oxford, UK

Preface

The importance of centrifugation is demonstrated by the presence of centrifuges in almost all departments of cell and molecular biology and many medical laboratories. Probably every worker in these laboratories has had to use one or more of the centrifugation techniques at sometime. Centrifugation is used to separate and purify many different types of biological particles in both research and routine applications. While it is often a case of simply following a well-documented, almost foolproof procedure, there are many occasions when a knowledge of the basic theory of centrifugation is useful, for example, in order to modify an existing technique or to design a new technique to suit the work in hand.

This book is aimed primarily at those completely new to centrifugation. It sets out to explain what centrifugation is, what it does, how it does it and the types of centrifuges, rotors, centrifugation techniques and density-gradient media available. Part 1 commences with a brief, simplified description of the theoretical aspects of centrifugation, followed by discussions of the machines, the methods and the media used in centrifugation.

The machines. Since the mid 1970s the development of centrifuges and rotors has moved ahead quickly, resulting in many changes to their design. The introduction of new drive systems and new materials together with microchip technology has revolutionized the production of the latest machines. We shall briefly describe these developments, the effects they have had upon separation techniques and the advantages they can give over older methods.

The methods. The more commonly used centrifugation techniques — differential pelleting, rate-zonal banding and isopycnic banding — are described in simple terms and the results that may be expected when using each technique are discussed.

The media. The properties of some of the media commonly used to provide density gradients are described, together with their effects upon biological material. The development of new media to enhance the separation and purity of particles is discussed.

v

In Part 2, the applications of the foregoing techniques are demonstrated by a selection of practical methods for the purification of each class of biological material — intact cells, subcellular organelles and macromolecules or macromolecular complexes — together with the types of centrifuges and rotors needed.

Apart from three simple, but important, equations, the mathematics of the dynamic behaviour of particles in a centrifugal field have been ignored, firstly because they are well explained in other works [see for example, *Centrifugal Separations in Molecular and Cell Biology*, 1978, (Eds G.D. Birnie and D. Rickwood), Butterworth] and secondly because they are not required for the practical, day to day uses of centrifugation.

While this book is mainly intended for those new to centrifugation techniques, it is hoped that it will also be of use to more experienced workers who may wish to develop a new separation method, or modify an existing one. Furthermore, many of the protocols in current use were developed before the advent of the latest centrifuges and newer density-gradient media. It is the hope of the authors, that the information provided here will be of assistance to those wishing to enhance and simplify older methods.

T.C. Ford
J.M. Graham

Contents

Safety Note

Centrifuges

Even low-speed centrifuges develop very high forces during rotation which are contained within the bowl in which the rotor operates. The bowl is lined with armoured steel for safety purposes. Providing operating instructions are followed, there is little danger of accidents causing injury. However, centrifuges and their rotors are very expensive items which can be damaged by careless usage. It is essential to balance the centrifuge tubes and their contents correctly before placing them into the rotor, and to ensure that the balanced tubes are placed in the buckets directly opposite each other. If this is not done, the rotor will be unbalanced and set up a vibration when rotating. Most centrifuges have a sensor which detects imbalance and stops the machine, but even so, great strain will have been put upon the drive-shaft and bearings, reducing the life of the machine. If the safety device fails, the rotor can tear apart like an exploding grenade, especially in high-speed and ultracentrifuges.

Centrifuge tubes

Centrifuge tubes come in many different sizes and are made of different types of material with different properties with regard to the centrifugation forces that they can withstand and their resistance to various chemicals. Using tubes under the wrong conditions can lead to them breaking in the rotor during centrifugation with several possible consequences: loss of a possibly important and costly sample; contamination of the centrifuge and surroundings by harmful substances contained in the sample material (radio-isotopes, pathogens, etc.); or imbalance in the rotor leading to the damage described above. Always clean buckets, rotors and the inside bowl of the centrifuge, especially after spillages. Always follow the makers operating instructions, use the correct tubes, correctly balanced and, if in any doubt, ask someone who knows; even ring the manufacturer for help.

1 A Theory of Centrifugation, Simplified

1.1 Sedimentation of particles under the influence of gravity

To start with a very simple example, we can look at what happens when we put a handful of mixed dry soil and sand into a tall glass of water. We will see that the particles immediately move towards the bottom of the glass, pulled down by the force of *gravity*. In this case the force is described as unit gravity or 1 *g*. After a short time, if we examine the material in the bottom of the glass, we will note that the mixture put into the top of the glass has separated into layers, with each layer consisting of particles of roughly the same size. Generally, the size of the particles in the sediment increases from the top to the bottom of the pellet (*Figure 1.1*). There will still be some particles suspended in the water moving slowly downwards, in some cases so slowly that we cannot see them move. Other particles may be seen right at the top of the water.

From our observations we can deduce that larger particles sediment (move toward the bottom) more quickly than do smaller particles and that very small particles sediment very slowly. However, some small, but very dense particles might sediment faster than larger but very light particles. We also note that some particles do not sediment, but remain floating at the top of the liquid. We have, in fact, just performed a very simple separation of the mixed particles due to the different sedimentation properties of the different particles. The separation was done under the influence of the Earth's gravity and is described as 'sedimentation at 1 *g*'.

1.2 Increasing the effect of gravity: the centrifuge

This, then, is the main principle of centrifugation; particles suspended in a fluid move under gravity towards the bottom of the vessel at a rate depending, in general, upon their size and density. Such sedimentation only occurs, of course, if the density of the fluid is less dense than the density of the particles. Centrifugation is a technique designed to utilize centrifugal forces which are greater than the force of gravity and thus speed up the sedimentation rate of particles. This is achieved by spinning the

1

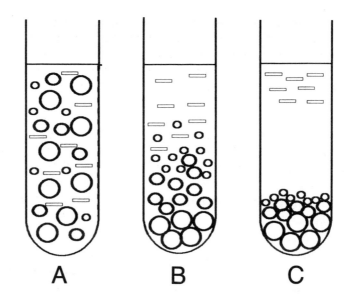

FIGURE 1.1: *The mixture of particles, initially throughout the tube (**A**) begins to sediment with larger particles sedimenting faster than the smaller (**B**). The very small particles sediment very slowly (**C**) and particles less dense than the medium remain at, or float to, the top.*

vessel containing the fluid and particles about an axis of rotation so that the particles experience a force moving them away from the axis, much in the same way as one experiences a force trying to throw one off a roundabout, or carousel. The machine which spins the vessels is called a *centrifuge* and the device that holds the vessels and rotates about the axis is called a *rotor*. The spinning of the rotor generates a force directed away from the axis; the force is measured in multiples of the Earth's gravitational force and is known as the *relative centrifugal field* (r.c.f.) or 'g force'. The relationship between the particles, medium and centrifugal field is shown in the following equation:

(1)
$$v = \frac{d^2(\rho_p - \rho_l)g}{18\mu}$$

where: v = velocity of sedimentation; d = diameter of the particle
ρ_p = density of the particle; ρ_l = density of the medium
μ = viscosity of the medium; g = r.c.f.

From the equation it can be seen that the sedimentation rate of a particle is proportional to its size and to the difference between the density of the particle and the density of the surrounding medium. When the particle density is equal to that of the medium, the sedimentation rate becomes zero, and when the medium density exceeds that of the particle, the sedimentation rate becomes negative; that is, its direction of migration is reversed and the particle will move towards the top of the centrifuge tube. The sedimentation rate will decrease as the viscosity of the medium increases, and will increase as the centrifugal force increases.

The relative centrifugal field (r.c.f. or g) generated by a rotor is dependent upon two factors: the speed of the rotor in revolutions per minute (r.p.m.) and the radius of rotation, that is, the distance from the axis of rotation. It can therefore be seen that particles further from the axis will experience a greater sedimenting force than those nearer the axis.

The following equations allow the r.c.f. to be calculated from a known r.p.m. and radius of rotation, and the r.p.m. to be calculated from a known r.c.f. and radius.

(2) $$\text{r.c.f.} = 11.18 \times r(\text{r.p.m.}/1000)^2$$

(3) $$\text{r.p.m.} = 299.07 \ \sqrt{\text{r.c.f.}/r}$$

where: r = radius in cm.

1.3 Density gradients

Simple pelleting procedures can achieve only a limited degree of resolution of biological particles, in order to obtain a higher degree of resolution, density gradients, or density barriers are required. A density gradient consists of a solution which increases in density from the top to the bottom of the centrifuge tube. It may be a continuous gradient, in which the density increases continuously, or discontinuous, in which the density increases in a series of discrete steps. The formation of density gradients will be described in a later chapter. A density barrier is a solution of a density designed to allow some particles to pass through it and others to be prevented from entering it.

2 Introduction to Centrifugation Techniques

The centrifugation techniques of main interest can be divided into three classes; *differential centrifugation*, *rate-zonal centrifugation*, and *isopycnic banding*.

2.1 **Differential centrifugation**

As in the water and soil experiment described in Section 1.1, the separations obtained depend upon the differences in the sedimentation rates of the various particles in the suspension. Instead of a mixture of sand and soil, a centrifuge tube is filled with a suspension containing a mixture of biological particles, such as mixed populations of intact cells or an homogenized tissue or organ. The homogenate will contain the contents of broken cells, consisting of nuclei, mitochondria, lysosomes and many other particles of different sizes. These will be distributed randomly throughout the medium in which they are suspended, usually a sucrose solution.

Like the sand and soil mixture, if the homogenate is allowed to stand long enough, the larger particles in the solution will sediment to the bottom of the tube. However, as these particles are much smaller and less dense than most of the sand and soil mixture, sedimentation will take much longer as the sedimenting force of gravity is opposed by the forces of diffusion and Brownian motion. For this reason, the sedimentation rate needs to be increased by use of a centrifuge. A brief description of the initial purification steps used to separate the components of a mouse-liver homogenate will demonstrate the purpose of differential centrifugation.

The homogenate is first centrifuged at 1000 g for 5–10 min, which is sufficient to pellet the nuclei; a consequence of their large size as compared with the other subcellular particles. However, in centrifuging fast enough and long enough to pellet all the nuclei, many of which started off close to the top of the solution, more slowly moving particles, which started nearer the bottom of the tube, will also be found in the pellet and therefore the pellet is far from consisting of pure nuclei. It is this that compromises the resolution obtainable by differential centrifugation. The supernatant is centrifuged at a higher centrifugation force to pellet particles of the next lower order of size and that supernatant again centrifuged at increasing time or speed in order to pellet successively smaller particles. Each pellet will, as in the case of the nuclear pellet, contain contaminating material, the final, microsomal pellet being the most pure of the pellets recovered, leaving a final supernatant containing the soluble fraction of the homogenate, macromolecules and salts, etc.

It is clear that the first pellet, the nuclear pellet, should have the highest percentage recovery, but also will contain the most contaminants. Reference to the equation for sedimentation velocity (*equation 1*, Section 1.2) shows that, for particles of equal density, a particle with a diameter of two units will sediment four times faster than one with a diameter of one unit. In such cases then, all the smaller particles in the lower quarter of the centrifuge tube would sediment in the time taken for the larger particles, initially at the top of the tube, to pellet. The successive pelleting of the supernatants will result in the later pellets having successively lower recovery rates due to a proportion of their potential material having been lost to the previous pellets. The microsomal pellet will therefore have the highest purity but the lowest yield. A proportion of the contaminants of any pellet can be removed by resuspending the pellet and repeating the centrifugation step several times, but this leads to a lower recovery of the material.

It can be seen, then, that differential centrifugation can be a useful method of obtaining partially purified fractions of the homogenate. To increase the purity of the fractions, one of the other techniques of centrifugation can be used, following partial purification by the differential pelleting method.

Pelleting is also used to concentrate material. For example, a 1 liter culture of bacteria may be centrifuged at about 2000 *g* for 15 min to pellet the cells and allow resuspension in a smaller, more concentrated, volume. The same applies to other material, with the centrifugation conditions adjusted to suit the particular particles.

2.2 **Rate-zonal centrifugation**

We have seen that differential centrifugation will partially purify fractions of the homogenate on the basis of size differences and sedimentation rates. Rate-zonal centrifugation operates on the same principles, applied in a different way.

The sample material, in our example one of the resuspended pellets from the liver homogenate, is loaded onto a preformed density gradient, usually sucrose. When centrifugation begins, the particles sediment through the gradient towards the bottom of the centrifuge tube as before. In contrast to differential pelleting, in which the sample material is distributed throughout the medium, the sample is initially present only on top of the gradient in a narrow band. As the material in the band moves down through the medium, the faster sedimenting particles move ahead of the slower, so that a number of bands or zones of particles of similar size are sedimenting at different rates. The distance between the zones increases with distance from the start position, thus increasing the *resolution* between zones. The speed at which the particles sediment is dependent, not only upon their size and to a smaller extent, density, but also upon the shape of the particles. The maximum density of the gradient must be less than the buoyant density of the particles. The need for a density gradient, and not just an homogeneous solution of an appropriate density, is twofold: firstly the gradient prevents convection within the tube, which would disrupt the bands of particles as they sediment; and secondly, the increase in density and viscosity of the gradient as particles move along the tube compensates for the increase in the relative centrifugal field, thus maintaining an almost constant rate of sedimentation of the particles.

In our example, the advantage of rate-zonal centrifugation over differential centrifugation is that, in the latter, as the sample is distributed throughout the centrifuge tube, the pellet is contaminated by other slower sedimenting particles. In the rate-zonal separation, all the particles start from approximately the same place, hence the need for the sample to be loaded as a narrow band on top of the medium. Because the difference in the r.c.f. between the top and bottom of the sample layer is very small, as compared to that in differential centrifugation, the faster sedimenting species are not contaminated by the slower, as they are when differential centrifugation is used. The method is, in fact, a little more complex than shown here since the density and gradient of the medium have to be chosen carefully, either from prior experience or by resorting to one of the computer simulation programs available (see Section 7.2), or even by trial and error.

So, both the rate-zonal and differential centrifugation methods rely mainly upon the effects of size and, to a lesser extent, density upon their sedimentation rates to separate the particles. However, particles, especially subcellular organelles, of similar function may have a wide range of different sizes, thus making rate-zonal separations less useful than one might expect. In addition, particles of similar size may be very different in other respects. For example, the various sub-populations of lymphocytes and monocytes found in blood, may be of similar size but are very different in function. To separate particles of similar size, but of different morphology and function, factors other than size must be considered. One such factor is the difference in the buoyant densities of the various particles. The buoyant density of a particle is its apparent density in a liquid medium and this is measured against the density of the medium. When the medium is of a density that allows a particle to remain stationary within it, neither sedimenting nor floating upwards, the density of such medium (ρ_l) is the same as the buoyant density of the particle (ρ_p) in that medium. The observed buoyant density of a particle depends upon the gradient medium used and this will be discussed in a later section of this book. Separation of particles on the basis of their buoyant densities requires the technique of *isopycnic banding* or equilibrium centrifugation.

2.3 Isopycnic banding

Isopycnic banding separates particles purely upon the basis of differences in their buoyant densities. This is achieved by loading the sample onto a density gradient already formed in the tube (a preformed gradient) or by mixing the sample with the gradient medium in a centrifuge tube and allowing a gradient to form during centrifugation (a self-forming gradient). The manner in which density gradients can be prepared is described in Section 5, 'Formation of density gradients'. When a preformed gradient is used, the sample can be loaded onto the top of the gradient, at the bottom, or at any other position within the tube. During centrifugation, the particles migrate to a point in the gradient where the density of the medium is equal to the buoyant density of the particles and at that point migration stops. This point is the isopycnic position of the particle. In this case, the size of the particle only affects the rate at which the particle reaches its isopycnic position. An important consideration, when using isopycnic centrifugation, is the viscosity of the medium. Higher viscosity generally leads to poorer resolution of the bands.

It may not be generally realized that particles loaded into the bottom of a gradient, first having been mixed with gradient medium to make the sample solution denser than the remainder of the gradient, can float upwards to find their own buoyant density. Therefore, bottom or middle-loaded samples, and samples initially distributed throughout the gradient, can migrate in either direction to their isopycnic position. This behaviour can be deduced from *equation 1* (Section 1.2), where the particle density, ρ_p, is less than ρ_l, and thus v becomes negative, indicating a reversal in the direction of migration. In flotation, as in sedimentation, the larger the particle, the faster it moves.

2.4 Conditions for centrifugation

Among the centrifugation techniques described above, differential centrifugation or differential pelleting is the most widely used and is probably the first technique to which anyone new to centrifugation will be introduced. Although it is the most simple procedure, it still has to be used with care if the optimum results are to be obtained. The limitations of differential pelleting with regard to the resolution of particles has been discussed, but consideration must also be given to the centrifugation force used, the time, the rotor and the medium in which the particles are centrifuged.

For pelleting procedures, fixed-angle rotors are generally most efficient and, when large volumes are involved, the fixed-angle rotors usually have a larger capacity. However, as can be seen from *Figure 3.1*, although the fixed-angle rotors have a shorter *pathlength* (see Section 5.3) than the swing-out type, the particles will hit the wall of the tube before sliding down to form the pellet. If the rotor is accelerated too fast, or if there is a high concentration of material in the tube, this can result in smearing the pelleting material along the length of the tube. When cells or other large particles are to be pelleted, this effect will be more pronounced and in the case of cells, can reduce recovery due to breakage and other damage. It is therefore advisable to use a swing-out rotor in such cases. When cells are pelleted, it is clear that those near the bottom of the tube initially will hit the bottom almost immediately, while those initially near the top will take longer. The early pelleting cells will therefore be subjected to a compressive force against the solid bottom of the tube and from the particles above them. Damage can thus ensue from these 'crushing forces' themselves and also, because the early pelleting cells will form a firmer pellet, resuspension from such a firm pellet will be more difficult and result in further damage.

One of the more serious problems encountered when trying to separate mixed cell populations on density gradients is that of cell aggregation. The tendency of cells to clump together in aggregates prevents a good separation of the different types of cell on continuous or discontinuous gradients and this is often seen as a corkscrewing rope of aggregated material along the length of the gradient which disrupts any bands present. In some cases this is due to damage which occurs during the washing and pelleting stages of the preparation and this can often be minimized by centrifuging the cells onto a density cushion of the medium to be used at the density gradient stage. In this way the cells do not suffer the severe compressive forces of pelleting as they are on a more elastic surface and, furthermore, resuspension of the cells is a much more gentle procedure.

While differential pelleting of cells is usually carried out in physiological saline or nutrient media, many other biological particles will require an inert, nonionic medium to be used. As will be discussed in the later section dealing with the properties of density-gradient media, the ionic strength of the medium in which particles are suspended can affect the conformation and/or function of macromolecular structures and organelles. Sucrose solutions, which are inert and nonionic, are commonly used in these instances.

3 *The Machines, Centrifuges and Rotors*

Centrifuges come in many shapes and sizes and with various degrees of technical sophistication. Apart from the specialized machines, covered briefly later in this book, they can be classified as: low-speed, high-speed and ultra-speed centrifuges. Advances in technology have allowed the introduction of direct drive systems, microchip processing for automated operation, the design of more compact models, the development of solid-state cooling systems, which eliminate the need for environmentally damaging CFCs, and more reliable performance. While all these advances mean that both internal and external features of centrifuges may change, the basic principle of a machine to increase the sedimentation rate of particles will remain.

The design of rotors has also advanced with the use of new materials to provide light-weight but strong models. While this has not led to great increases in the r.c.f. generated, it has reduced the stress imposed on the drive systems.

3.1 Low-speed centrifuges

These machines have maximum rotor speeds of less than 10 000 r.p.m. and vary from small, bench-top to large, floor-standing centrifuges. Low-speed centrifuges do not require the rotors to be run in vacuum, but it is useful if they have a temperature control mechanism. Most modern machines have sensors which act to cut off the drive if any imbalance is detected when the rotor is spinning. Some of the older models do not and this can result in serious damage to the machine if it is not immediately switched off manually as soon as vibration starts. The operating controls may consist merely of an on/off switch and a speed control, or they may include more sophisticated options such as, timers, temperature control, brake/ acceleration control and be able to convert the speed in r.p.m. to a readout of the r.c.f. generated.

Low-speed centrifuges are used to harvest and purify chemical precipitates, intact cells (animal, plant and some micro-organisms), nuclei, large mitochondria and the larger plasma-membrane fragments.

3.2 **High-speed centrifuges**

Broadly speaking, these machines are capable of producing rotor speeds of up to 21 000 r.p.m., but this classification on the basis of speed is no longer very well defined. New species of bench-top centrifuges are becoming popular which now straddle the low-speed/high-speed categories, with maximum speeds of 6000–20 000 r.p.m. (depending upon the centrifuge). Moreover, at the time of writing, advances in design are also making the division between a high-speed and an ultra-speed centrifuge difficult to define. Until recently, high-speed machines could be considered to be those with top speeds of about 20 000 r.p.m., generating r.c.fs of 40–50 000 g. Now, however, there are centrifuges described as 'high-speed' or 'super-speed', which are capable of providing r.c.fs of up to 120 000 g at rotor speeds of up to 30 000 r.p.m., although only the smallest volume rotors are capable of running at the top speeds and producing the r.c.fs advertised.

As with the low-speed range, the degree of sophistication varies greatly with the capabilities of the machine. Microchip technology has introduced computerized programming of the machines, making some of the display boards resemble the flight-deck of Concorde, but after the first shock at the size of the 'Operating Instructions' handbook, the object of the exercise is to help the operator and make results highly reproducible.

The high-speed centrifuges require refrigeration systems to regulate temperature and the higher speed machines must have vacuum systems, all of which add to the cost of the finished article. The versatility of any particular machine will depend upon the rotor systems compatible with it, whether they are rotors made by the centrifuge manufacturer for that machine or those of other makers which are suitable for it.

High-speed centrifuges are used to isolate and purify viruses, mitochondria, chloroplasts, lysosomes, peroxisomes, intact Golgi membranes and small membrane fragments.

Improvements in the capabilities of high-speed centrifuges mean that many separations which used to require an ultra-speed centrifuge, can now be carried out using one of the new high-speed or super-speed machines (see Section 9). This offers some advantage in these cost-conscious days, as high-speed machines are less expensive than the ultracentrifuges and can serve as general purpose machines for many different separation techniques.

3.3 **Ultracentrifuges**

The remarks regarding the high-speed centrifuges in regard to technology and advanced designs, apply equally to the ultra-speed ranges. Here, we will consider ultracentrifuges to be machines capable of over 30 000 r.p.m., in fact, some can reach up to 120 000 r.p.m. and develop over 600 000 g, with the appropriate rotor. The introduction of direct-drive motors means there are fewer moving parts to wear out and the use of carbon-fibre material in the design of rotors has resulted in lighter rotors and thus less stress on the drives, contributing towards more reliable operation and less maintenance.

Ultracentrifuges are used to isolate and purify all membrane vesicles derived from the plasma membrane, endoplasmic reticulum and Golgi membrane, endosomes, ribosomes, ribosomal sub-units, plasmids, DNA, RNA and proteins.

3.4 Capacity of centrifuges

In terms of purely dimensional considerations, the floor-standing models can generally accommodate larger sample volumes than the bench-top models.

Low-speed centrifuges of the large, floor-standing type, will accept up to 6 × 1 liter, while the medium size models will take up to 4 × 500 ml samples and the bench-top models 4 × 250 ml.

Floor-standing high-speed centrifuges normally have a maximum capacity of 6 × 500 ml and bench-top models can accommodate 8 × 15 ml or 4 × 50, with some able to take 6 × 100 ml. The majority of the bench-top models are aimed at the small volume sample (1–2 ml).

Floor-standing ultracentrifuges have a maximum capacity of 6 × 100 ml, with the larger volume rotors having a severe restriction on the maximum rotor-speed, sometimes as low as 20 000 r.p.m.. The bench-top ultracentrifuges are specifically designed for small (1–4 ml) samples.

Each centrifuge will offer a number of rotor capacity specifications, the sample number increasing with decreasing sample volume. For example, a centrifuge which has a maximum capacity of say, 6 × 500 ml, may also accommodate 8 × 50 ml and 64 × 1 ml samples. Appendix B gives detailed specifications of a large selection of centrifuges.

3.5 Temperature control

3.5.1 Operational requirements

Generally speaking, intact cells are centrifuged in low-speed centrifuges at room temperature; this is particularly the case if centrifugation is part of a continuous cell culture process. As most mammalian cells are grown at 37°C and bacteria often at room temperature, it serves no useful purpose to carry out centrifugation at 4°C, indeed, it may actually be detrimental to mammalian cells. Low temperature centrifugation is only necessary when any subsequent analysis requires the cell metabolism to be arrested or the cells are to be used for the preparation of subcellular organelles or macromolecules.

3.5.2 *Centrifuge requirements*

All ultracentrifuges and large-capacity high-speed and low-speed centrifuges are refrigerated, while small-capacity bench-top low-speed and high-speed machines offer refrigeration as an option; although the trend is for all centrifuges to be refrigerated. Temperatures can usually be controlled over the range -10 to $+40°C$.

In low or high-speed bench-top centrifuges, the heat generated by friction of the spinning rotor and by the drive itself, can be fairly effectively dissipated to the environment. However, operation of these machines at maximum speeds, particularly for long times (over about 20 min), can lead to warming of the sample. In all other centrifuges, refrigeration is needed to overcome the greater heat generation. Rotors of ultracentrifuges are always run *in vacuo* to reduce frictional heating, which is difficult to dissipate by refrigeration alone at speeds above 20 000 r.p.m. The new high-speed centrifuge produced by Sorvall is capable of up to 28 000 r.p.m. with the fixed-angle rotors, using the vacuum operation at the higher end of the speed range. Swinging-bucket type rotors generate more frictional heat than do angle rotors unless they are wind-shielded.

3.6 **Rotors**

The design of most centrifuges allows the drive system to accept rotors of different sizes and capacities according to the separation to be carried out. However, there has been a trend in recent years for single rotors, particularly those of the low- and high-speed varieties, to be capable of accepting a large range of tube sizes and types through the use of adaptors. This reduces the necessity of purchasing a large range of different rotors. The rotors have to hold the centrifuge tubes or bottles containing the samples or gradients to be centrifuged, such vessels varying in volume from about 0.5 ml up to 1000 ml.

The rotors have three basic designs: swinging-bucket, in which the sample tubes are carried in buckets able to swing outwards to a horizontal position as the rotor spins; fixed-angle rotors, in which the sample tubes are held at a fixed angle to the vertical position; and the vertical-tube rotors (which are themselves, fixed-angle), in which the tubes are fixed in the vertical position. The basic design of each type is shown in *Figure 3.1*, but there is a great deal of variation in the capacities of the different rotors available.

In general, low-speed centrifuges use the swinging-bucket type rotors and have interchangeable rotor heads to accept buckets of different capacities. For example, one head may take buckets with spaces for a large number of small tubes or a small number of large tubes, while another

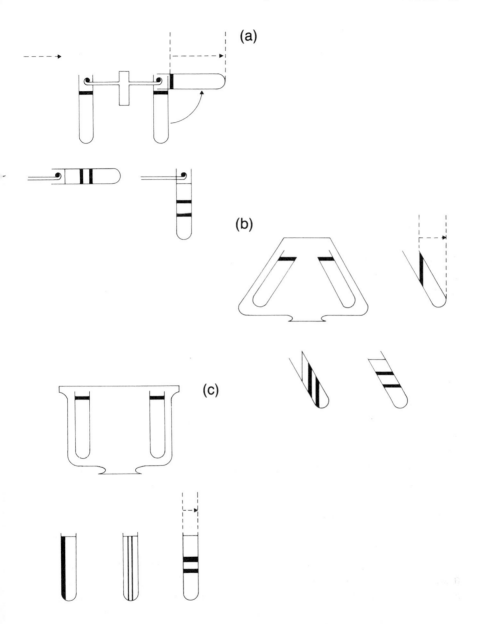

FIGURE 3.1: *Types of rotor. (**a**) Swing-out rotor: the gradient does not reorientate relative to the centrifuge tube. The pathlength is the length of the tube. (**b**) Fixed-angle rotor: The gradient has to reorientate relative to the centrifuge tube. The pathlength is less than the length of a tube. (**c**) Vertical-tube rotor: the gradient has to reorientate relative to the tube. The pathlength is the diameter of the tube.*

head may take buckets which accept 500 or 1000 ml bottles. Some machines do accept fixed-angle rotors, but these tend to be for small volumes and have a higher speed range. These machines can be adapted to many tasks requiring low-speed centrifugation. Due to the relatively low forces generated, the rotors do not require the same high degree of precision engineering and high-stress materials as those designed for the higher speed machines and thus they cost significantly less.

Rotors for the high-speed and ultra-speed centrifuges are subjected to high stresses and, at the high rotational speeds used, precise balance of the rotors and the sample tubes they contain is absolutely essential. While balance is important for all rotors, those used at high speed will not tolerate the rough and careless handling sometimes given to the low-speed versions. Chips, dents or corrosion will lead to imbalances and subsequent damage to the drive systems, even though there are sensors to detect such imbalance and cut the drive. The swinging-bucket rotors of the high and ultra-speed centrifuges have each bucket and often its cap numbered, as well as the place to which it fits on the rotor. These positions must be strictly observed. In all rotors, the centrifuge tubes used must be balanced to within the limits set by the manufacturer, usually within 0.1 g in the case of high and ultra-speed models. Swinging-bucket rotors must always be used with all buckets attached, even if all do not contain sample tubes. When only one sample gradient is being run, it is, of course, necessary to have a balancing tube in the opposite bucket. While with low-speed centrifugation it is only necessary to balance the tubes by weight, with high-speed runs the tubes should be loaded in the same manner. That is, if a density gradient is formed in one tube, the balance tube should contain a similar gradient.

In many cases, at forces of less than 50 000 g, partially filled sample tubes can be used, while at higher speeds it is safer, and in many cases essential, that they are completely full to prevent tubes collapsing under the high centrifugal fields. This is most important when using the fixed-angle or vertical-tube rotors.

Rotor designation, such as 6 × 14 ml, indicates a six-place rotor with a maximum capacity of 14 ml for each tube, although in most cases, the actual volume of the tube depends upon the wall thickness of the tube and the useful volume is usually 5–10% less than that quoted. Details of rotor and tube capacities are given in Appendix C.

3.7 Centrifuge tubes

For the sedimentation of intact cells, it is necessary to use the minimum force required to obtain a pellet which is not too compact. A very compact pellet of cells leads to cell damage and loss of function or recovery. It is therefore advisable to use a conical-bottomed centrifuge tube in a swinging-bucket rotor, as the conical-shaped bottom retains the pellet more effectively as the supernatant is removed. The pellet is also more securely held against the liquid vortex, which may occur if the rotor slows from maximum speed to rest too quickly.

All tubes for high-speed and ultracentrifuge rotors are round-bottomed, except for the small-volume Eppendorf tubes (up to 1.5 ml), which fit a number of high-speed rotors.

Which particular tube material is chosen is a matter of personal preference in many cases, but there are factors to be considered relating to the possible reactivity of the tube material and, of course, the maximum centrifugation field which it is designed to withstand.

3.7.1 *Glass centrifuge tubes*

Ordinary Pyrex® tubes can withstand forces of around 2000 g, while Corex® tubes can be centrifuged at up to 12 000 g. They are still commonly used for all low-speed centrifugation, but most such work can be carried out in plastic tubes. Plastic tubes have the advantage of cheapness and can be obtained ready sterilized. The main disadvantage of some types is their relative opacity compared with glass.

3.7.2 *Solvent resistance and sterilization*

Glass is the most obvious choice for solvent resistance but plastic tubes do have some restricted solvent resistance. In general, polycarbonate is least resistant and polyallomer most resistant to organic solvents. Of the synthetic materials, only cellulose acetate cannot be autoclaved, but this is one of the few synthetics that can be UV sterilized. In the authors' experience, repeated autoclaving of polycarbonate tubes leads to stress damage, which in turn may lead to breakage during centrifugation. The manufacturers provide extensive information with regard to solvent, salt and pH resistance, and provide sterilization procedures for their centrifuge tube materials.

Polycarbonate and polyallomer tubes are the most commonly used materials for high-speed and ultracentrifuges, with polypropylene being rather less common. Polycarbonate tends to be more expensive, but is very transparent and thick-walled tubes of this material are the only ultracentrifuge tubes that can be used partially filled. If material is to be harvested by puncturing tube walls, polyallomer tubes should be used, although Beckman produces a tube called 'Ultraclear®', which has the transparency of polycarbonate but can be punctured for harvesting banded material.

3.7.3 *Tube caps*

Since most ultracentrifuge tubes have to be full, they require caps to prevent spillage during centrifugation. The caps can be of varying degrees of complexity and also serve to support the tube to prevent them collapsing in high centrifugal fields. Modern sealed tubes provide the most effective containment, shaped rather like bottles with rounded bases, their tops are either mechanically or thermically sealed. These tubes are used to advantage in vertical-tube rotors and/or with potentially hazardous samples.

3.8 Safety practices when using the centrifuge

Operational errors which are serious enough to cause the centrifuge to malfunction are all associated with careless or uninformed handling of the sample tubes or rotor, rather than with the actual operation of the centrifuge. The only error which may arise in the latter, is setting a rotational speed greater than that permitted for the particular rotor in use. However, most modern centrifuge rotors are coded so as to prevent such a setting being accepted by the machine. Earlier models will accept and run the rotor until its speed limit is reached, then register an overspeed error and either cut the motor or continue the run at the acceptable rotor speed. There are also self-limiting rotors designed to be of sufficient mass so that the drive system is not capable of running them in excess of the maximum stipulated speed.

3.8.1 *Preparing centrifuge tubes and samples*
Centrifuge tubes for the low- and high-speed rotors can be used partially filled regardless of the material from which they are made and the centrifugal field used (provided, of course, they are suitable for that centrifuge). The main problem is overfilling tubes when using fixed-angle rotors. Uncapped tubes should only be filled, at most, up to the recommended level remembering that the meniscus reorients in the tube during centrifugation and so overfilling will result in spillage.

Failure to fill ultracentrifuge tubes adequately and/or to secure their caps, where fitted, will cause the tubes to collapse under high centrifugal fields. Cap failures can usually be attributed to lack of routine inspection and cleaning or faulty assembly. *Note* that some tubes are restricted in the maximum r.c.f. to which they can be exposed.

3.8.2 *Balancing the rotor*
Rotors must be in balance before spinning begins and this means carefully balancing diametrically opposed tubes. The material in the two tubes must be of the same volume and density: a gradient in one tube must be balanced by an identical gradient in the opposite tube. It is the position of the centre of gravity in the opposing tubes that is most important, not just the weight and volume. Different pairs of opposing tubes in the same rotor can contain different sample material and gradient profiles, so long as the opposing tubes of each pair are identical. All buckets must be in place on a

swinging-bucket rotor, even if all do not contain gradients or samples, and each numbered bucket must be fitted with its correct cap and be fitted into its correct, numbered position on the rotor.

Rotors may have their stipulated maximum speed *derated*. The object of derating rotors is to reduce the maximum speed at which they were initially rated since, under certain conditions, running at the maximum rated speed might cause severe stress to the rotor, thus leading to rotor failure. Derating is an essential practice when running solutions which will form very high-density gradients. Caesium chloride solutions of >1.2 g/ml initially, will self-generate gradients which have a very high density at the bottom of the tube, overstressing the rotor in high centrifugal fields. In such conditions the rotor must be derated, that is, run at a lower maximum speed than that inscribed upon it. The manufacturer's literature will indicate the conditions that call for derating and the degree of derating needed. Derating is also recommended for rotors that have run for more than a specified number of hours, which is the reason for keeping a log-book for each high-speed and ultracentrifuge rotor. Again, follow the manufacturer's recommendations.

3.8.3 *Cleaning*

Simple cleaning procedures should be routinely used for rotors and tube caps. Washing in warm water containing a nonionic detergent such as Teepol®, followed by rinsing in distilled water and air-drying, should be the usual practice after use, and always done if spillage has occurred. Aluminium rotors are especially prone to corrosion and should *never* be used with strong salt gradients. All caesium chloride gradients should be centrifuged in titanium rotors.

Special attention should be paid to 'O' rings and screw-threads on rotors and caps. Examine them for deterioration and, in the case of the 'O' rings of rotor lids, they must be *lightly* smeared with vacuum grease. The 'O' rings are important seals divorcing the interior of the rotor and sample tubes from the vacuum in the centrifuge chamber. Screw-threads should be free of grit or debris which might restrict easy turning.

Adherence to these simple rules and an appreciation of the problems that might arise will help prevent rotor failure and the consequent heavy repair bills.

Modern centrifuges are designed to contain any rotor failure within the rotor chamber. A rotor coming loose from the drive-shaft at high rotational speeds has a huge amount of energy to dissipate, in the course of which the inside of the rotor chamber is ripped to pieces and the rotor, especially a swing-out type, will be broken into small pieces. A thick, armoured wall around the chamber ensures that major accidents where containment fails, are now very rare. Vibration sensors detect any imbalance and shut off the drive, but if any excessive vibration is noticed and the drive is still running, switch off manually at once.

Lid-locking devices are fitted to most modern machines to prevent the lid being opened when the rotor is spinning, but some older low-speed and a few high-speed machines do not have this device and the lid can be opened before the rotor stops. *Never* do this, and never try to increase the braking of the rotor by applying manual pressure. This is exceedingly dangerous and also, the sudden change in deceleration rate will create a vortex in the centrifugation medium, upsetting the pellet or bands of material present.

3.8.4 *Use of potentially pathogenic or toxic material*

So long as the rotor is sealed properly, any spillage will be contained within it. If spillage is released from the rotor in a modern centrifuge it should be contained in the sealed rotor chamber. Never use a bench-top, non-refrigerated low- or high-speed centrifuge for such material since these centrifuges are cooled by drawing air in from, and expelling air out to, the laboratory environment, thus creating aerosols of any spillages.

Using an appropriate decontaminant, the rotor should be washed out and the chamber swabbed down. For biological material, SDS, ethylene dioxide or ethanol are permissible for the decontamination of rotors.

For work which merits the use of a Containment Facility, centrifugation must be carried out within that facility. If highly pathogenic material is used, a code of practice for centrifugation work should be devised which is approved by the appropriate national Health and Safety Authority.

3.9 **Problems**

Except for the sedimentation of cells (see Section 3.5.1), the centrifugation of most biological material is carried out at 4°C. Ideally the centrifuge chamber should be pre-cooled to this temperature. In many centrifuges this can be done without the rotor being in the chamber, or while the rotor is stationary, but some older ultracentrifuges need the rotor to be spinning at low speed within the chamber. The rotor itself should also be pre-cooled, so it is best to cool rotor and chamber together. If no pre-cooling is carried out, it may take up to 1 h or longer for temperature equilibrium to occur.

Temperature sensors in centrifuge chambers, at best, register the temperature at the surface of the rotor; at worst, the atmosphere within the chamber — not the temperature within the tube. If a rotor is not run *in vacuo*, then the higher the speed of rotation, the more likely it is that the temperature in the tube is higher than the set temperature. Each high-speed Sorvall rotor has a temperature calibration graph relating the set temperature to the tube temperature at different rotational speeds.

Temperature will also affect the actual rate of sedimentation of a particle. According to *equation 1* (Section 1.2), v is inversely proportional to μ, the viscosity of the medium; moreover, viscosity, particularly of sucrose and polysaccharides, such as Ficoll®, is highly temperature dependent *(Table 3.1)*. The rate of sedimentation of a particle through 50% (w/w) sucrose at 0°C is about half that at 10°C.

Lists of a number of commercially available centrifuges and rotors are given in Appendices B and C, together with their specifications.

TABLE 3.1: *The viscosity of sucrose solutions: dependency upon temperature and concentration.*

Sucrose	Viscosity (N.m^{-2}.s × 10^{-3})				
% (w/w)	0°C	5°C	10°C	15°C	20°C
20	3.7	3.1	2.6	2.5	1.9
30	6.7	5.4	4.5	3.8	3.2
40	14.6	11.5	9.2	7.5	6.2
50	44.7	33.2	25.2	19.5	15.4

4 *The Media*

4.1 **Choosing a suitable density-gradient medium**

In the years since the introduction of centrifugation as a technique for the separation of biological particles, many different compounds have been used to form density gradients and enhance the separations. In separating the heterogeneous mixture of particles in biological sample material, the main purpose is to purify a particular type of particle for intensive study, to deduce such things as its structure, function and its position within the interacting components of its *in vivo* surroundings. It follows, therefore, that if the results of such studies are to be meaningful, the particles studied must be maintained as closely as possible to their *in vivo* structural and functional properties. While it must always be borne in mind that the very separation of particles from their natural environment is likely to alter their structure and function to a lesser or greater degree, every effort must be made to minimize such changes. Centrifugation itself, which acts to increase the gravitational force experienced by all matter, may have adverse effects upon the material investigated. This can be minimized by reducing the time and r.c.fs employed to the minimum required and by ensuring that the gradient medium in which the biological particles are suspended is innocuous to the particles. The choice of the density-gradient medium is therefore often vital for successful purifications.

4.2 **Main classes of biological material**

In general, the types of biological particles to be separated fall into three main categories:

(1) intact single-cell suspensions (e.g. peripheral blood, dissociated tissue);
(2) subcellular organelles (components of cell homogenates; e.g. mitochondria, lysosomes and membrane-bound vesicles);
(3) macromolecules and macromolecular complexes (nucleic acids, proteins and nucleoprotein complexes).

Each of these classes of material requires different centrifugation techniques, different centrifugation conditions and close attention to the gradient medium employed.

4.2.1 *Class 1: Intact single-cell suspensions*

Suspensions of cells may be obtained from body fluids such as peripheral blood, amniotic fluid or semen; be prepared from dissociated whole tissues; or be suspensions of cultured cells. Suspensions from body fluids or dissociated tissues will contain cells of different types, sizes and function, while those of cell cultures often contain cells of a single type, although during culture the cells may be undergoing differentiation involving changes in structure and function. In all cases, separating the cells into their various categories by centrifugation requires care in the choice of medium and the centrifugation technique to be used.

In their natural or cultured environment the cells will be subject to closely controlled conditions of temperature, pH and osmotic pressure, and so removal from these conditions may have adverse effects upon the cells. During the separation and purification procedures it is necessary to maintain the preferred environment as closely as possible or to return the cells to that environment as soon as possible after purification.

It is because of the need to maintain the environment of the cells during centrifugation, that the choice of a medium to be used for density gradients is so important. The compound itself should be inert and non-toxic to the cells, be capable of forming a gradient with a density range sufficient to band the cells at their isopycnic positions and be able to maintain an osmolality close to that of the cells throughout the gradient. The osmolality of the gradient in relation to that of the cells is important in regulating the buoyant density of the cells for isopycnic banding. When subjected to an environment of lower osmolality (*hypo*tonic), cells (and other membrane-bound particles) tend to take in water and swell, sometimes to the extent that it causes them to burst. Under conditions of high osmolality (*hyper*tonic), the cells tend to lose water and shrink. Such variation in the size of the cells with the gain or loss of water, decreases or increases their buoyant density and therefore affects the position at which they will band in the gradient. Sometimes these effects can be used to advantage to alter the buoyant density of particular cells selectively in order to enhance a separation.

As cells are relatively large particles they can sediment either to form a pellet or to reach their isopycnic position in a short time using low centrifugation speeds, typically about 800 g for 30 min or less. This means that they are not subjected to very high gravitational fields or long centrifugation times and preformed gradients are necessary. Details of specific conditions for the separation of some cell types will be described later.

4.2.2 *Class 2: Subcellular organelles*

Suspensions of subcellular organelles are usually obtained initially from the homogenization of cells or tissues. Some starting material (particularly tissue culture cells) and some isolation procedures (liver plasma membrane) use hypo-osmotic homogenization medium. It is good practice to make the homogenate iso-osmotic (approx 9% sucrose) as soon as possible. Because of the complexity of an homogenate, the choice of medium in which to carry out the centrifugation is often a compromise. Divalent cations (Mg^{2+} and Ca^{2+}) are beneficial in maintaining the integrity of nuclei, but they are detrimental to the recovery of respiratory control in mitochondria and can cause membrane aggregation. The size of membrane fragments and vesicles and hence their behaviour during centrifugation will depend on the homogenization medium and homogenization strategy. The homogenate will contain particles varying in size from the comparatively large nuclei (3–12 μm) to the smallest vesicles and microsomes (0.05–0.2 μm) and finally, the soluble components such as macromolecules and salts.

Generally, one, or possibly two, components of the homogenate are required to be purified at any one time, for example, the mitochondria or lysosomes, or particular domains of the plasma membranes. The preparation of the homogenate will depend upon the organelle(s) to be finally purified.

The range of centrifugation forces required for the isolation of subcellular organelles in iso-osmotic (0.25 M) sucrose is from about 1000 *g* (5–10 min) for nuclei and membrane sheets to 100 000 *g* (30–60 min) for small membrane vesicles. In the homogenate, each type of organelle exists in a range of sizes so that a specific organelle will sediment at a variety of r.c.fs. For example, most mitochodria sediment at 10 000 *g* for 10 min, but some at 3000 *g* for 5 min.

The composition of the media used for the homogenisation and for the subsequent centrifugation steps is chosen to maintain the integrity of the organelle to be purified. The density-gradient medium must not be damaging to the organelle and therefore the criteria applied to that used for the cell separations must again be considered. In the case of membrane-bound particles, the effect of osmolality upon the buoyant density of the organelles will affect the resolution of bands of the different organelles. However, sucrose gradients are commonly used and because of their high viscosity, r.c.fs many times those needed to pellet the organelles are required for equilibrium banding (100 000 *g* is often used for purifying nuclei). The conditions for the separation of specific organelles will be discussed later.

4.2.3 *Class 3: Macromolecules and macromolecular complexes*

The very small size of even the largest of these types of particle, means that their separation by centrifugation must require the use of an ultracentrifuge and very long centrifugation times (up to 60 h in some cases). However, recent developments of super-speed machines and new gradient media have allowed such high forces and prolonged times to be reduced in some cases. While the very high gravitational fields, applied to these particles for long periods, may not have an adverse effect upon their structure or function, the medium in which they are centrifuged may very well affect their structure and therefore, their function. The nature of the gradient medium can affect the structure of macromolecules by denaturing them, or causing conformational changes, it can also alter the buoyant density of the molecules by changing their degree of hydration, which depends upon the amount of free water in the solution. Traditionally, gradients of heavy metal salts have been used for the fractionation of nucleic acids.

4.3 Properties of the compounds used as centrifugation media

Many different compounds have been used as centrifugation media for the purification of biological material and generally, they can be classified as:

(1) salts of the alkali metals, e.g. caesium chloride;
(2) low molecular weight sugars, e.g. sucrose;
(3) high molecular weight polysaccharides, e.g. Ficoll® or dextran;
(4) colloidal silica suspensions, e.g. Ludox®, Percoll®;
(5) the iodinated compounds (a) ionic, e.g. sodium metrizoate; (b) non-ionic, e.g. Nycodenz®.

The above classes have all been found useful for particular separations but often have serious disadvantages for others. When planning a new separation method, or wishing to modify a current technique, it is necessary to know the properties of the gradient medium to be used and to evaluate its effect upon the biological material in order to optimize the results obtained.

4.3.1 *Solutions of caesium salts*

Among the salts of the various alkali metals, caesium chloride and caesium sulfate are probably the most widely used as centrifugation media. Solutions with densities of up to 1.9 and 2.01 g/ml respectively can be prepared, but these solutions are of high ionic strength and can damage some biological material, e.g. dissociation of unfixed nucleoprotein complexes, such as chromatin and ribosomes. While the solutions of these salts have very low viscosities, they do have very high osmolalities, making them totally unsuitable for banding osmotically sensitive particles, such as cells. The available 'free' water molecules in a solution, that is, water molecules not bound to the solute, is an important factor and defines the term 'water activity' of a solution — there is low water activity where there are few free water molecules and high water activity when there are many. The importance of the water activity of a solution used as a density-gradient medium is seen in the effect it has upon the degree of hydration of biological particles exposed to it. In solutions of low water activity, biological particles, such as DNA, tend to become partially dehydrated, while in solutions of high activity they will be closer to the fully hydrated state. This, in turn, affects the conformation and the buoyant density of the particle exposed to that particular solution. In the case of DNA, its buoyant density in caesium chloride gradients is observed to be about 1.6–1.7 g/ml, while in solutions with higher water activities, such as Nycodenz solutions, it bands at a density of about 1.12 g/ml. Solutions of the salts of alkali metals form relatively steep gradients, which reduces their resolving power, but centrifugation on self-forming caesium chloride gradients is used to separate supercoiled, linear and relaxed-circular DNA and is especially useful for the preparation of purified plasmid DNA. It is often useful to look at the effect of cations and anions binding the molecule and obviously ionic media, such as solutions of caesium salts, cannot be used for such studies. In these cases a nonionic medium is required to allow carefully monitored aliquots of different ions to be added to the medium in order to observe their effects.

Caesium chloride solutions are not sufficiently dense to band RNA, but caesium sulfate gradients can be used for that purpose, banding DNA, RNA and protein. High molecular weight RNA may, however, aggregate and precipitate in the presence of sulfate ions, which also cause precipitation in liquid scintillation fluids.

Other salts of alkali metals have been used as centrifugation media and details of their use may be found elsewhere [1].

4.3.2 *Sucrose solutions*
Sucrose solutions are very widely used for centrifugation purposes, being relatively cheap and simple to prepare and providing nonionic solutions which are inert to most biological material. At concentrations above 9.0% (w/v) they are hyperosmotic with regard to most mammalian cells and, as at that concentration the density is only 1.03 g/ml, the banding of cells and osmotically sensitive particles in isotonic conditions is not possible. The high viscosity of dense sucrose solutions also leads to poor resolution in the isopycnic banding of particles. Sucrose solutions can be prepared with densities of up to 1.39 g/ml, but at such high concentrations they have low water activities, causing partial dehydration of macromolecules. As a consequence, the buoyant densities of macromolecules in sucrose gradients are greater than the maximum densities obtainable with these solutions

and therefore they are not used to band macromolecules isopycnically. There have been reports of damage to the respiratory control mechanisms of mitochondria and disruption of vesicular structures by sucrose [2,3] and a number of common assay procedures for nucleic acids and proteins cannot be carried out in the presence of sucrose.

In spite of these deficiencies, sucrose gradients have been usefully applied to the separation of subcellular organelles, viruses and in some cases, cells, although the high osmolality of sucrose solutions leads to shrinkage of osmotically sensitive particles and thus the buoyant densities of organelles and vesicles are higher than those found under iso-osmotic conditions. This often causes co-banding of particles, which in iso-osmotic environments show significant differences in their buoyant densities.

Sucrose solutions are the most commonly used gradients for the rate-zonal separation of particles, for example, the separation of the sub-units of ribosomes. Computer programs are available to simulate conditions of centrifugation allowing the correct parameters to be chosen for the separation required (see Section 7.2).

4.3.3 *Polysaccharides*
Because of the high osmolality of sucrose solutions, polysaccharides, such as the naturally occurring glycogen and dextran and, more recently, the synthetic polysaccharide, Ficoll, have been used as density-gradient media. While for any given concentration, the polysaccharides have lower osmolalities than the mono- and disaccharide solutions, their viscosities are very much higher, leading to poorer resolution of bands and longer centrifugation times. Polysaccharides cannot be removed from the sample material by dialysis or ultrafiltration, but the biological particles can be recovered by diluting the gradient fractions and pelleting. Gradients of polysaccharides have been used for the rate-zonal and isopycnic banding of cells and organelles.

4.3.4 *Colloidal silica media*
The colloidal silicas, which, as the name implies, are colloidal suspensions of silica particles, not true solutions, were first recognized as possible density-gradient media in the early 1960s [4,5]. The original, unmodified silica particles, however, were toxic to cells and unstable in salt solutions. The stability was found to be increased by coating the particles with polymers such as dextran, polyethylene glycol (PEG) or polyvinylpyrrolidone (PVP), which also decreased their toxicity [6,7]. The colloidal silica suspensions provide solutions of very low osmolalities, but with relatively high viscosities as compared with caesium salts or the iodinated media.

Probably the most widely known and used of the colloidal media to date, is that marketed under the name of Percoll. Percoll consists of PVP-coated silica particles of 15–30 nm diameter, supplied as a colloidal suspension with a density of 1.130 g/ml, which is essentially free of unbound PVP and non-toxic to cells. The low osmolality of Percoll makes possible the formation of isotonic gradients by the addition of appropriate concentrations of salts or sucrose, and the colloidal nature of Percoll allows the rapid formation of self-generated density gradients. The suspension is destabilized in the presence of salts or sucrose and thus cannot be autoclaved after the addition of salt or sucrose solutions to provide stock working solutions from the Percoll solution supplied. Therefore all Percoll diluents have to be autoclaved separately. Some cells tend to phagocytose the silica particles, which can cause problems and removal of the medium from the biological material can be difficult, involving a number of washes and consequent loss of sample material. The silica particles tend to pellet during high-speed centrifugation and, when trying to pellet particles smaller than the silica particles themselves, the silica will pellet before the biological particle. Nevertheless, Percoll can be used for a wide range of successful applications in the field of cell biology.

4.3.5 *Iodinated media*

The iodinated media, derived from metrizoic acid, were originally developed for use as X-ray contrast media, but their properties which permitted the preparation of dense solutions at relatively low concentrations were soon recognized as possible density-gradient media. The early derivatives, such as sodium metrizoate and diatrizoate, formed ionic solutions but still found very useful and widely used applications, especially in the form of Lymphoprep® [8], a solution of sodium metrizoate and Ficoll (a synthetic polysaccharide added to aggregate the erythrocytes and increase their sedimentation rate), which has been used widely for many years for the separation of mononuclear blood cells from erythrocytes and granulocytes. Further development of the X-ray contrast media resulted in the production of metrizamide [9], the first of the nonionic, iodinated media early in the 1970s and later, the second of such compounds, Nycodenz [10]. The nonionic media, again derivatives of metrizoic acid, are able to provide dense solutions at relatively low concentrations and with low osmolalities and viscosities. Extensive studies have found them to be inert, non-toxic and to form stable gradients. They can be used to prepare isotonic gradients for the separation of cells and membrane-bound particles and, if necessary, the medium can easily be removed from sample material in the gradient fractions. Nycodenz, the later of the two nonionic media, has several advantages over metrizamide in that it can be autoclaved and also, the absence of a ribose sugar molecule in the case of Nycodenz, allows many of the standard chemical assays to be carried out in its presence. However, in general the two compounds are very similar and apart from the differences given above, a separation technique describing the use of one is likely to be easily applied to the other. The nonionic, iodinated media are very versatile in their applications over a wide range of biological particles, especially in regard to the purification of cells and membrane-bound particles.

4.4 Summary of the properties of density-gradient media

Table 4.1 indicates some of the common uses of the different density-gradient media. The selection of a suitable medium will depend upon the type of material to be separated and the subsequent treatment of the purified fractions. It can be seen that solutions of the salts of alkali metals or sucrose, while very useful, have limited ranges of applications, while the high viscosity of solutions of the polysaccharides also limits their uses. Among the colloidal silicas, Percoll has been most widely used and reported over a wide range of separation procedures. The iodinated media, especially the nonionic compounds metrizamide and Nycodenz, have been been used for separations and purifications in all the classes of biological material and are very versatile.

TABLE 4.1: Separations possible using the different classes of gradient media available.

	Nucleic acids	Protein	Nucleo-protein	Cell organelles	Intact cells	Viruses
CsCl	+	+	some	−	−	+
Sucrose	−	−	some	+	−	+
Polysaccharide	−	−	−	+	+	+
Percoll	−	−	−	+	+	+
Nycodenz	+	+	+	+	+	+

References

1. Spragg, S.P. (1978) in Centrifugal Separations in Molecular and Cell Biology (G.D. Birnie and D. Rickwood, eds). Butterworth, London, p. 7.

2. Zimmer, G., Keith, A.D. and Packer, L. (1972) Arch Biochem. Biophys., **152**, 105.

3. Kurokawa, M., Sakamoto, T. and Kato, M. (1965) Biochem. J., **97**, 833.

4. Mateyko, G.M. and Kopac, M.J. (1963) Ann. N. Y. Acad. Sci., **105**, 185.

5. Pertoft, H. and Laurent, T.C. (1968) in Modern Separation Methods of Macromolecules and Particles (T. Gerritsen, ed.). Wiley Interscience, New York, Vol. 2, p. 71.

6. Pertoft, H. and Laurent, T.C. (1977) in *Methods of Cell Separation* (N. Catsimpoolas, ed.). Plenum Press, New York, Vol. 1, p.25.

7. Pertoft, H. and Laurent, T.C. (1978) *Anal. Biochem.*, **88**, 271.

8. Bøyum, A. (1968) *J. Clin. Lab. Invest.*, **21** Suppl. 97, 1.

9. Rickwood, D., Hell, A., Birnie, G.D. and Gilhus-Moe, C.C. (1974) *Biochim. Biophys. Acta*, **342**, 367.

10. Rickwood, D., Ford, T.C. and Graham, J.M. (1982) *Anal. Biochem.*, **123**, 23.

5 Formation of Density Gradients

5.1 Types of density gradient

Density gradients may be *discontinuous*, that is, with sharp interfaces between the layers of different densities, or *continuous*, in which the density gradient increases gradually throughout the length of the centrifuge tube. The differences can be likened to the difference between walking up steps or walking up a hill. In the same way that a hill can be steep or have a gradual slope, so a density gradient may be steep or shallow. A graph drawn to illustrate the increase in density along the length of a centrifuge tube will show the steepness or profile of a gradient (see *Figure 5.1*). The shape of the profile may be linear, concave or convex and this can be pre-determined by the manner in which the gradient is prepared and the medium used to form the gradient. The profile of a gradient can be important in that it can determine the *resolution*, i.e. the degree of separation, of the bands of material to be purified. Given two types of particle, one with a buoyant density of 1.10 g/ml and one with a buoyant density of 1.11 g/ml, and two continuous gradients of equal volumes, one with a density range of 1.0–1.2 g/ml and the other with a density range of 1.08–1.12 g/ml, the two bands of particles will have a greater distance separating them on the second gradient, i.e. they have a higher degree of resolution. This is shown graphically in *Figure 5.1*.

5.1.1 *Discontinuous gradients (step gradients)*
These are formed by carefully layering solutions of increasing densities into a centrifuge tube, so that sharp interfaces are formed between the different solutions. This is best carried out using a syringe of suitable capacity with a long blunt needle or metal cannula with a bore of about 0.7–1.5 mm. The aliquots of different densities may be underlayered, i.e. the least dense solution is put into the tube first and the successively denser solutions introduced to the bottom of the tube to underlay the previous solution (the recommended method), or by loading the most dense solution first and carefully overlaying the less dense solutions. The needle must always be full of solution when underlayering so that no air bubbles are pushed through to disturb the gradient being formed.

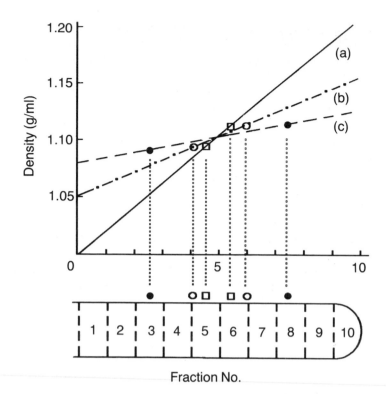

FIGURE 5.1: *Resolution of gradients. The resolution, or the degree of separation of the bands, depends upon the density profile of the gradient. Three density profiles are shown:* **(a)** *1.0–1.2 g/ml (————);* **(b)** *1.05–1.15 g/ml (–·–·–);* **(c)** *1.08–1.12 g/ml (– – –). The densities of the particles of the two bands are 1.09 g/ml and 1.11 g/ml. A tube is shown with the gradient fractionated into 10 fractions. Profile (a) would have the bands in adjoining fractions (□). Profile (b) has bands better resolved (○). Profile (c) has the bands much farther apart (●). Note: The buoyant density of each band is the same on all gradients.*

5.1.2 *Continuous gradients*

These may be formed in several ways. When a discontinuous gradient has been made up as described above, some gradient media, such as the iodinated media or sucrose, allow a gradient to be formed by diffusion. If the centrifuge tube containing such a discontinuous gradient is allowed to stand overnight (about 16 h) at a temperature between 4 and 8°C, a continuous gradient will be formed by diffusion. Alternatively, the process may be speeded by sealing the top of the tube and laying it in a horizontal position at room temperature (about 18–25°C) for 45–60 min. This increases the interfacial areas and thus allows the diffusion process to form the gradient more quickly. Other methods involve the use of mechanical gradient-mixers, which mix two solutions of different densities to form a continuous density gradient in the centrifuge tube. *Figures 5.2* and *5.3* illustrate two such machines. The gradients described above are called *preformed gradients* for the obvious reason that they are formed prior to centrifugation.

Continuous density gradients may also be formed by centrifuging a solution, which initially is of the same density throughout the centrifuge tube, for sufficient time and at a sufficient speed. During centrifugation, a gradient is formed as the density decreases at the top of the tube and increases toward the bottom. The time, speed and temperature of centrifugation for a gradient of a particular profile to form depends upon the properties of the solution used as the gradient medium and the type of rotor used. This type of gradient is known as a *self-forming gradient*, but cannot be formed with all types of gradient media. The solutions of caesium salts, Percoll and the iodinated media may be used; but fast diffusing solutions, such as sucrose, cannot be used since diffusion of the solute negates the sedimenting forces which form the gradient. Percoll, because of its colloidal nature, can provide gradients formed *in situ* much faster than the other media.

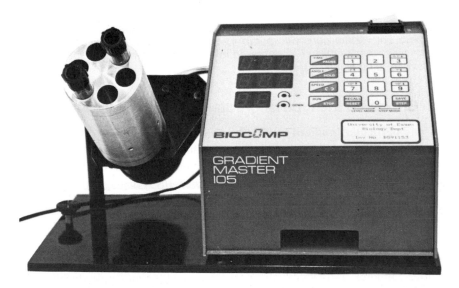

FIGURE 5.2: Density gradient machine. The dense and light solutions are layered in the centrifuge tubes. A gradient is generated by computer-controlled mixing: the tubes are rotated at a set speed and angle for a set time.

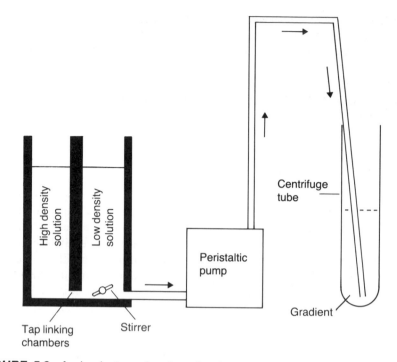

FIGURE 5.3: *A simple two-chamber density-gradient mixer. The low density solution is in the chamber with outlet to pump. When the tap is opened, the high density solution enters the low density chamber and the solutions are mixed as they pass out via the peristaltic pump to the bottom of the centrifuge tube. A progressively denser mixture is delivered to the tube, uplifting the lighter solution and forming a continuous gradient.*

5.2 Choosing the type of gradient required

For any particular gradient separation, a choice of gradient type must be made. In the simplest terms, preformed gradients are used when the size of the particles to be purified are large enough to reach their isopycnic position, i.e. the position in the gradient where the density of the medium is equal to the buoyant density of the particles, in less time than it takes a gradient to self-form. This is very important in the case of large particles such as cells, cell organelles and viruses. The use of preformed gradients means that shorter centrifugation times and lower r.c.fs can be used and thus delicate structures are not subjected to excessive forces for long periods. Smaller particles, such as macromolecules and macromolecular complexes, require much longer centrifugation times to band them, during which a gradient can be self-generated and it is usual to mix the sample with the gradient medium and centrifuge for the required time in self-forming gradients.

When separating osmotically sensitive particles such as cells and some organelles, in addition to the necessity of preformed gradients to reduce centrifugation time, it is also advantageous to use gradients which are iso-osmotic with the cells in order to prevent damage that may be caused by osmotic stress. The separation of particles with significant differences in size and/or density, can often be carried out using a single solution of the appropriate density. Centrifugation time and/or the r.c.f. can be reduced by the use of media of low viscosity.

5.3 Choosing a rotor

The choice of rotor will depend upon the use to which it is being put and the gradient profile to be generated. While all rotors generate centrifugal fields, the r.c.f. obtainable is related to the speed at which the rotor turns and the maximum distance from the axis of rotation. Thus, for a given speed in r.p.m., the maximum force generated depends upon the radius of rotation. As discussed previously, the rotors may be of fixed-angle, vertical-tube or swinging-bucket type and their different characteristics are, to a great extent, due to differences in their *pathlengths*. The pathlength refers to the distance along the axis of rotation that a particle can travel during centrifugation. Referring to *Figure 3.1*, it can be seen that in the case of a swinging-bucket rotor, the possible pathlength is the distance from the top to the bottom of the centrifuge tube, due to the bucket-swinging outwards to a horizontal position during centrifugation. The fixed-angle rotors, however, have a pathlength determined by the diameter of the centrifuge tube used and the angle at which the tube is held in the rotor. The diagrams of *Figure 3.1* show that, with equal tube size, the pathlength of particles centrifuged in a vertical-tube rotor will be shorter than if centrifuged in a rotor with tubes angled at, for example, 30°. The angle at which the rotor holds the tubes during centrifugation and the differences in pathlength must be considered when choosing the rotor for a particular task. If one wishes to pellet material, as in the case of differential centrifugation techniques, the most efficient rotor to use is a fixed-angle rotor, since these have a relatively short pathlength compared with a swinging bucket rotor of the same tube size. The particles are sedimented to the bottom of the tube, forming a pellet to one side. Particles pellet rapidly in a vertical rotor, but in this case, the pellet forms on the side of the tube, not at the bottom, and therefore the material of the pellet tends to contaminate the supernatant to a greater extent when the rotor stops because it is in contact with the whole length of the tube.

For the preparation of self-forming gradients, the shorter pathlengths of the fixed-angle and vertical rotors allow the gradients to form much more quickly than in swinging-bucket rotors and are the rotors of choice when banding particles such as macromolecules which take longer to reach their equilibrium position than it takes the gradients to self-form in the fixed-angle rotors. The angle of the rotor, the temperature during centrifugation, the time of centrifugation and the concentration of the medium in the centrifugation tube will all have an effect upon the shape of the gradient generated. Generally, the vertical-tube rotors and those in which the tubes are held at a small angle to the vertical produce gradients which are linear throughout most of the sedimentation path, while the large-angle and swing-out rotors produce gradients which have a shallow profile in the centre, but are steep at the top and bottom of the tube. As mentioned earlier, the shape of the gradient profile affects the resolution of the bands. The relative times taken for gradients to form under the same centrifugation conditions depends upon the medium used. The colloidal silicas will generate gradients very rapidly as compared with the iodinated media or caesium chloride solutions and thus self-formed gradients can be prepared quickly for particles which can reach their isopycnic positions in a short time.

Generally, when such large particles as cells and the larger organelles are to be separated, preformed gradients are used and swinging-bucket rotors are more suitable. Moreover, centrifugation in swing-out, rather than fixed-angle rotors, limits the changes in the profile of the preformed gradient to the top and bottom of the tube, leaving the centre position relatively undisturbed (*Figure 5.4*).

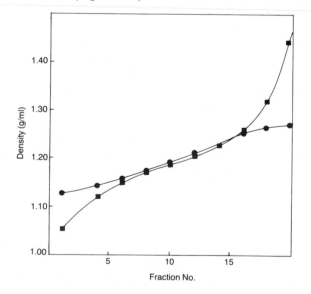

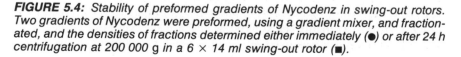

FIGURE 5.4: *Stability of preformed gradients of Nycodenz in swing-out rotors. Two gradients of Nycodenz were preformed, using a gradient mixer, and fractionated, and the densities of fractions determined either immediately (●) or after 24 h centrifugation at 200 000 g in a 6 × 14 ml swing-out rotor (■).*

5.4 Formation of iso-osmotic gradients

When cells and other osmotically sensitive particles are to be separated on density gradients, it is obviously advantageous to be able to maintain a very narrow osmotic range throughout the gradient, close to that of the cells' natural environment. Not only does this prevent damage to the cells due to osmotic stress, it also ensures that the buoyant density of the cells does not vary as they pass through the gradient. Changes in buoyant density occur when the cells are subjected to changes in the osmolality of the surrounding medium, losing water and shrinking in *hyper*-osmotic conditions and gaining water and swelling in *hypo*-osmotic environments, respectively increasing or decreasing their buoyant densities. Deliberate manipulation of the osmolalities of gradient solutions can sometimes be useful to enhance the separation of cell species of closely similar buoyant densities in iso-osmotic conditions. In order to form iso-osmotic gradients for the separation of biological particles, the criteria discussed in Section 4 must be considered when choosing the gradient medium to be used. Here we describe the formation of iso-osmotic density gradients using either Nycodenz or Percoll solutions, which have both been described extensively for this purpose in recent years.

Most mammalian cells will band in iso-osmotic gradients at buoyant densities of less than 1.15 g/ml and the natural osmotic environment of such cells is 280–330 mOsm. Both Nycodenz and Percoll can be used to cover the density range required while maintaining osmotic profiles close to that of the cells. The osmolality of 30% (w/v) Nycodenz solution in water is 300 mOsm at a density of 1.159 g/ml, while Percoll is supplied as a suspension with an osmolality of 25 mOsm, at a density of about 1.13 g/ml. The osmolality of human serum is close to 290 mOsm and that of mouse serum about 320 mOsm, with sera of many other mammalian species within this range. Gradients can therefore be prepared at densities sufficient to band such cells, which will maintain an osmolality throughout the gradient compatible with that of the cells. The examples of gradients given here apply to the osmotic environment of human cells, but adjustments can be made where necessary to suit other conditions. The osmolality of solutions is increased by the addition of other solutes and therefore the addition of buffers and salts, etc. necessary for the well-being of particles during centrifugation, increases the osmolality of the solutions.

A density gradient is formed by the increase in concentration of the medium used from the top to the bottom of the centrifuge tube. If a gradient is formed of 10–30% (w/v) Nycodenz made up in water, the osmolality will increase from about 100 mOsm (hypotonic) at the top of the tube to 300 mOsm at the bottom (close to isotonicity with human serum). Centrifuging cells from the top to bottom of such gradient would subject them to considerable osmotic stress as they pass first from their natural environment

TABLE 5.1: *The effect of the osmolality of diluent solutions upon the osmotic profile of the gradient.*

Fraction number	Osmolality of gradient fractions		
	Gradient (a)	Gradient (b)	Gradient (c)
1	95	270	276
2	109	271	276
3	118	271	281
4	130	277	283
5	144	282	287
6	169	293	292
7	204	300	301
8	235	303	302
9	258	302	302
10	288	304	298
11	293	298	295
12	293	300	297

Three gradients of 7–28% (w/v) Nycodenz were prepared using the Gradient Master (*Figure 5.2*). The 7% Nycodenz solution was diluted from the 28% stock solution by addition of (a) water, (b) 0.75% NaCl solution, or (c) 4.1% glucose solution. Each gradient was fractionated into 12 fractions and the osmolality of each fraction measured.

into a very *hypo*-osmotic region which causes them to take up water from the external environment and thus swell. The swelling can be to such an extent that the cells are lysed, or their buoyant density is decreased to an extent that prevents migration into the gradient medium. The cells that do migrate through the gradient are subjected to increasing osmolality causing them to lose water and shrink. These changes in volume cause changes in the buoyant densities of the cells, thus affecting their isopycnic banding positions and also their sedimentation rates. In order to maintain a stable osmotic environment throughout a gradient, an osmotic balancer has to be added to the solutions. *Table 5.1* shows the osmotic profiles of three identical density gradients, prepared in the gradient-mixer depicted in *Figure 5.2*. The gradients were prepared by loading 6 ml aliquots of 7% (w/v) Nycodenz solution under-layered with 6 ml of 28% (w/v) Nycodenz solution. The 28% solution was made up in water and had an osmolality of 294 mOsm. The dilutions of the 28% solution to 7% (w/v) Nycodenz were made by adding one part of the 28% solution to three parts of (a) water, osmolality 0 mOsm, (b) 0.75% NaCl, 242 mOsm, or (c) 4.1% glucose solution, 245 mOsm. This gave final osmolalities as (a) 86 mOsm, (b) 267 mOsm, and (c) 271 mOsm. The three gradients were mixed at the same time in the Gradient Master, for 1.45 min at an angle of 75° and at 20 r.p.m. The gradients were fractionated into 12 × 1 ml fractions and the osmolality of each fraction was measured.

When two solutions of different densities are layered into a tube, diffusion occurs across the interface seeking to equalize the concentrations of solutes in the two solutions. Each solute in the solutions has its own diffusion rate which occurs independently of the other solutes and the final result of diffusion would be equal concentrations of all solutes throughout the two initial solutions. In the iso-osmotic gradients to be described, the density through the gradient is mainly contributed by the concentration of the medium, Nycodenz or Percoll, and the osmotic profile is evened out by the addition of salt or sugar solutions.

5.4.1 *Iso-osmotic Nycodenz gradients*
A 27.6% (w/v) solution of Nycodenz containing 5 mM Tris or Tricine buffer together with millimolar amounts of other salts as required, has a density of 1.146 g/ml and an osmolality of about 290 mOsm (depending upon the concentrations of salts, etc. added). This stock solution can be diluted by a diluent solution containing the same concentration of salts and buffer as the stock solution, plus an osmotic balancer. The osmotic balancer may be sodium chloride, sucrose or glucose solution, or cell-culture medium diluted to a suitable osmotic strength. One might expect that when two solutions of the same osmolality are mixed, the osmolality remains constant, however, this is not necessarily so. When the stock Nycodenz solution is mixed with an equal volume of a diluent solution of equal osmolality, the resulting solution has an osmolality of 320–330 mOsm. To provide an osmolality of 290 mOsm in the final solution, the diluent needs to be about 250 mOsm. Such diluents can be made up in the buffer solutions described by the addition of 0.75% (w/v) sodium chloride, 7.45% (w/v) sucrose or 4.1% (w/v) glucose, or by diluting cell-culture medium to an osmolality of 250 mOsm. The stock Nycodenz solution can be diluted with any of these diluents as shown in *Table 5.2*.

TABLE 5.2: *Dilutions of stock 27.6% (w/v) Nycodenz, using 0.75% NaCl or 7.45% sucrose solutions as diluent.*

Nycodenz % (w/v)	Dilution ratio (Nyco:dil)	Osmolality ±3 mOsm	Density (g/ml)	
			0.75% NaCl	7.45% sucrose
27.6	1:0	290	1.146	1.146
18.4	2:1	295	1.098	1.105
13.8	1:1	290	1.075	1.086
9.2	1:2	280	1.050	1.066

The diffusion rate of sodium chloride is greater than that of Nycodenz so in the time that it takes for a linear gradient of Nycodenz concentration to form through the medium, the faster diffusion of the salt produces a flatter profile, the final result being that the osmolality is rather lower at the top of the gradient than at the bottom, as compared with the starting values. Thus the final osmotic profile of a 10–27.6% Nycodenz gradient, using the sodium chloride diluent, varies from about 280 mOsm at the top of the gradient to about 310 mOsm at the bottom, (*Figure 5.5*). These differences in the osmotic environment from top to bottom are not sufficient to cause significant stress to cells but will have some effect upon their buoyant densities, thus affecting their banding positions. As will be described in Part 2, such changes can, in some cases, be advantageous.

If four equal volumes of the dilutions described (say 2 ml aliquots in a 10 ml tube) are underlayered into a centrifuge tube (lowest concentration first), diffusion will start to occur across the several interfaces. It is clear

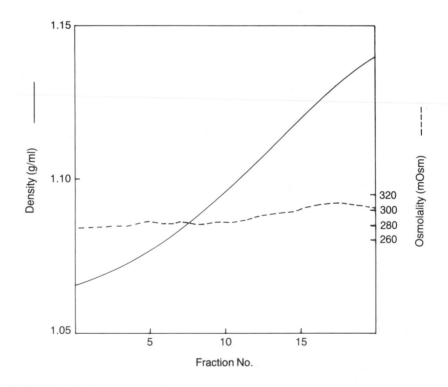

FIGURE 5.5: *Osmotic profile of a density gradient formed by the diffusion method. A four-step Nycodenz gradient, 10–27.6% (w/v), using the 4.1% glucose solution as diluent, was allowed to diffuse to form a continuous gradient. The gradient was fractionated and the density (———) and osmolality (– – – –) of each fraction was determined.*

that, initially, the Nycodenz concentration is highest at the bottom of the tube and successively lower in the dilutions toward the top, while the reverse is true for the salt or sugar concentrations (*Table 5.3*). The net diffusion of Nycodenz is therefore in the opposite direction to that of the components of the diluent solutions.

TABLE 5.3: *Initial salt concentration in diluted aliquots of 28% (w/v) Nycodenz solution, using 0.9%, 0.75% and 0.6% (w/v) NaCl solutions as diluents.*

Nycodenz % (w/v)	NaCl % (w/v)		
	0.9	0.75	0.6
28	0	0	0
21	0.225	0.187	0.150
4	0.450	0.375	0.300
7	0.675	0.562	0.450
4	0.771	0.643	0.510

The table shows that in discontinuous gradients formed from the 4–28% Nycodenz dilutions depicted, the Nycodenz will diffuse towards the bottom of the column, while the salt will diffuse towards the top of its column, in order to try and equalize the respective concentrations throughout the solutions.

Using the sucrose or glucose diluents, the slower diffusion rates of the solutes are closer to that of Nycodenz, which results in a narrower osmotic range, about 5–10 mOsm between the top and bottom of the gradient. The short centrifugation times and low r.c.fs used to band cells and osmotically sensitive particles will have a negligible effect upon the density and osmotic profiles of the gradients.

When preparing density gradients of different ranges, e.g. 5–20% (w/v) Nycodenz, some adjustment to the diluent solutions will be necessary to maintain the osmotic profile within the narrow limits described, as can be seen by reference to *Table 5.2*. Diluting the stock solution to a final concentration of 5% (w/v) Nycodenz with one of the diluents above, will result in an osmolality of about 270 mOsm which, upon diffusion, will have an osmolality of about 265 mOsm at the top of the gradient. Therefore, when diluting the stock solution to concentrations of less than 10% (w/v) Nycodenz, the osmolality of the diluent should be about 280 mOsm. *Table 5.4* shows a range of Nycodenz concentrations and the concentration of salts or sucrose required to obtain osmolalities close to 290 mOsm. However, in many cases, the effects of small osmotic differences are not great enough to damage the cells and where an adequate separation can be obtained without such precise adjustment, it may be omitted. In all work

TABLE 5.4: *Shows the effects of NaCl diluents of different osmotic strengths upon the osmolality of diluted fractions of a stock Nycodenz solution.*

Nycodenz concentration % (w/v)	Dilution ratio (Nyco:NaCl)	Osmolality (mOsm) using NaCl diluents		
		0.90%	0.75%	0.60%
21	3:1	312	294	279
14	1:1	309	289	260
7	1:3	304	270	231
4	1:6	298	258	215

Stock Nycodenz solution 28% (w/v) in water: 1.148 g/ml, 294 mOsm, refractive index 1.3790.

Osmolality of NaCl diluent solutions: 0.9% NaCl, 287 mOsm; 0.75% NaCl, 242 mOsm; 0.6% NaCl, 193 mOsm.

Gradients formed by the diffusion method, with four aliquots diluted from the stock solution using the 0.75% NaCl diluents, i.e. 21%, 14%, 7% and 4% (w/v) final Nycodenz concentrations, will have an osmotic profile after diffusion, varying from about 260 to 310 mOsm, due to the higher salt concentration at the 4% end diffusing toward the low salt end at a faster rate than the Nycodenz diffuses in the other direction. This effect can be minimized when using dilutions of Nycodenz of less than 10% (w/v) Nycodenz, by using a higher salt diluent, for example, 0.9% NaCl.

undertaken, it is sensible only to make preparations as complicated as is necessary to obtain the desired results.

5.4.2 *Iso-osmotic Percoll gradients*

Forming iso-osmotic gradients with Percoll requires a slightly different approach. Percoll is supplied as a colloidal suspension at a density of 1.13 g/ml and an osmolality of 25 mOsm. The very low osmolality of Percoll means that the osmolality of its gradients is effectively determined by the diluent solution used. Unlike true solutions, the colloidal silica particles of Percoll and other colloidal silica media, contribute to the density, but not the osmolality of the medium, the 25 mOsm osmolality of Percoll is therefore due to soluble factors in the medium. Concentrations of 0.15 M NaCl and 0.25 M sucrose are close to 290 mOsm thus close to isotonicity with mammalian cells. Diluting the initial Percoll suspension with 1.5 M NaCl, 2.5 M sucrose or a 10× concentrated culture medium, in the ratio of 9 parts Percoll to 1 part diluent provides a stock solution of about 300 mOsm. Further dilution of the stock with isotonic salt, sucrose or medium will maintain the tonicity of the diluted solutions. When density gradients are formed, either using a gradient mixer or a short period of centrifugation, the same concentration of the osmotic balancer throughout the tube means that no net diffusion of solutes occurs to affect the osmolality, which should therefore remain almost constant throughout the tube. The relatively short times and speeds used for the isopycnic banding of cells will only have a minor effect upon the preformed gradient profile, although, as the silica particles have faster gradient-forming properties than those of true solutions, the changes in gradient profiles will be greater in colloidal media than in solutions. Centrifugation in swing-out, rather than fixed-angle rotors, limits the changes in the gradient profile to the top and bottom of the tube, leaving the centre portion of the gradient relatively undisturbed (*Figure 5.4*).

The densities of 1.5 M and 0.15 M NaCl solutions are 1.058 g/ml and 1.0058 g/ml respectively, and those of 2.5 M and 0.25 M sucrose, 1.316 g/ml and 1.0316 g/ml, therefore the choice of diluent will have an effect upon the properties of the gradients. Sucrose diluent will create denser gradients than those made using sodium chloride, and the higher viscosity of sucrose solutions will affect the sedimentation rates of both the silica particles and the sample particles.

If we ignore for the moment the volume of liquid displaced by the silica particles and only consider the total volumes involved, it can be seen that

TABLE 5.5: Dilution factors for Percoll.

Density of Percoll solution, 1.13 g/ml (see package label for exact density)
Osmolality of Percoll solution <25 mOsm
Density of 1.5 M NaCl = 1.058 g/ml
Density of 2.5 M sucrose = 1.316 g/ml

To 9 parts Percoll solution add 1 part NaCl or sucrose solution to prepare a stock Percoll solution.

Using 1.5 M NaCl: density of stock solution 1.123 g/ml, osmolality 295 mOsm
Using 2.5 M sucrose: density of stock solution 1.149 g/ml, osmolality 290 mOsm

Stock solution is diluted to the required densities by addition of 0.15 M NaCl, density 1.0058 g/ml or 0.25 M sucrose, density 1.0316 g/ml.

$$Dvol = Pvol \ \frac{(\rho_P - \rho)}{(\rho - \rho_D)}$$

$Dvol$ = volume of diluent
$Pvol$ = volume of stock Percoll
ρ_P = density of Percoll stock
ρ_D = density of diluent
ρ = density required for diluted stock Percoll

---

preparing the stock Percoll suspension from the initial medium supplied, will result in media of different densities, depending upon whether sucrose or saline solutions are used as the diluent. The addition of 1.5 M NaCl solution to Percoll in the ratio 10 parts Percoll to 1 part NaCl, will result in a final density of this stock solution of 1.123 g/ml, while using the 2.5 M sucrose solution in the same ratio, the final density of the stock solution will be 1.149 g/ml. The osmolality of both stocks will be closely similar, about 300 mOsm. Self-generated gradients using sodium chloride or culture medium as diluent will form 2–3 times faster than those using sucrose diluent. *Table 5.5* shows the range of densities and osmolalities obtained using different dilution factors and different diluent solutions.

Nycomed Pharma of Oslo, Norway, and Pharmacia-LKB of Uppsala, Sweden, supply comprehensive booklets on the properties and applications of Nycodenz and Percoll respectively.

6 *Fractionation and Analysis of Gradients*

6.1 **Fractionation of gradients**

After the preparation and centrifugation of the gradients, the purified material must be removed for subsequent work or study. When discontinuous gradients, whether single-step or multiple-step have been used, the fractions of interest are usually found at the density interfaces and can be harvested using a Pasteur pipette or syringe fitted with a metal cannula. On continuous gradients, sometimes visible bands are apparent, well separated from each other, and these may be harvested in the same manner. Often, however, the regions of different particles are either not clearly visible and/or are present as overlapping bands and it is then necessary to fractionate the whole gradient and analyse the contents of each fraction. Several methods are available for such fractionations, some of which are now described.

A very simple method is to puncture the bottom of the centrifuge tube with a needle, with the tube held in an upright position, and allow the gradient to drip out, drop by drop. This can be an effective method but can cause cross-contamination of the different fractions, especially if there is pelleted material present, as material from the pellet will contaminate all the fractions passing through it. In cases where banded material tends to adhere to the sides of the tube, it will contaminate fractions from above the band as they move down.

Gradients can by fractionated by introducing a slim cannula to the bottom of the tube and using a peristaltic pump to draw out the gradient from the bottom to be collected in the desired fraction volumes. Again, pelleted material will cause problems and sticky material on the sides of the tube may cause cross-contamination. The problem caused by material from the banded particles adhering to the sides being scoured off to contaminate fractions from above as they are drawn down past the adhering material, can be minimized by fractionating slowly, allowing any particles scoured from the sides the opportunity to sink towards its own fraction, and also minimizing the chances of it being scoured from the sides in the first place. Pumping from the bottom, however, is not a recommended method as the resolution of the gradient fractions is compromised due to a reversal of the density gradient as the fractions are drawn into the cannula so that dense solution is above less dense. As the less dense solution tries to float upwards through the dense solution, the resolution is thus decreased.

The most easily controlled method for fractionating whole gradients is probably by using 'upward displacement'. This involves use of an apparatus designed to introduce a dense fluid to the bottom of the centrifuge tube to lift the gradient upwards into a collection device that allows the gradient to be harvested from the top (*Figure 6.1*).

The fluid introduced to the bottom of the centrifuge tube must, of course, be denser than the densest part of the gradient. Dense sucrose solutions have been successfully used for this purpose, but the use of a liquid such as Maxidens®, a fluorocarbon solution with a density of

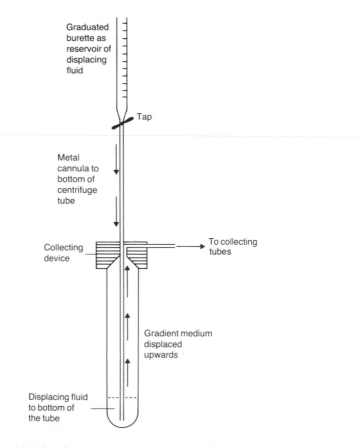

Graduated
burette as
reservoir of
displacing
fluid

Tap

Metal
cannula to
bottom of
centrifuge
tube

Collecting
device

To collecting
tubes

Gradient medium
displaced
upwards

Displacing fluid
to bottom of
the tube

FIGURE 6.1: *Fractionation by upward displacement. Displacing fluid is introduced from the graduated burette via a metal cannula to the bottom of the centrifuge tube. The gradient medium is displaced upwards into the collecting cap and then passed into the chosen fraction-collecting tubes.*

1.9 g/ml is recommended. Maxidens is immiscible with aqueous solutions and so does not contaminate gradient fractions. It has a low viscosity, making it easier to handle than dense sucrose solutions and is autoclavable for sterile work and can be recovered for re-use. As with other methods of fractionating the whole gradient, the presence of pelleted material and material scoured from the sides of the tube may present problems. However, using the upward displacement method does not lead to contamination of other fractions by pelleted material if a little care is taken.

When the displacing fluid is introduced to the bottom of the tube, do not allow the cannula to disturb any pellet, if present, but place the tip of the cannula a fraction above the pellet and allow a very slow flow of the displacing fluid to settle on the pellet and so displace the rest of the gradient upwards. Some or all of the pellet may float to the top of the displacing fluid but, if the operation is carried out carefully, will not swirl up to mix with the gradient above. Take care not to allow the tip of the cannula to penetrate the pellet as this can cause a temporary blockage which releases suddenly, swirling contaminating material up into the gradient medium, or block the cannula completely. The scouring of material adhering to the sides of the tubes can be minimized, as before, by fractionating slowly. The gradient is displaced upwards and passes into the collecting device fitted to the top of the centrifuge tube, then being carried to a rack of test-tubes or other containers set up to receive the gradient fractions. *Figure 6.1* shows one such device diagrammatically. The conical section of the collecting device enhances the resolution of the gradient.

6.2 Analysis of gradient fractions

If the behaviour and identification of a particular fraction is well established, either because the method itself is well documented, or it is a standard procedure yielding consistently reproducible results, then fractionation and analysis of the whole gradient will probably not be necessary. If, for example, a mixed population of intact cells are centrifuged and banded, it is often sufficient to remove the visible bands and examine them microscopically to ascertain purity and the presence of any contaminating species. With cell homogenates, however, there will be a great variety of particles present, with different banding densities and sedimentation rates.

Although it is possible to obtain partially purified fractions using differential pelleting before preparing a continuous or discontinuous gradient, a great deal of contaminating material is still almost certain to be present.

Let us illustrate the need for careful analysis of gradient fractions by considering the 'relatively simple' purification of mitochondria. If a pellet of partially purified mitochondria, prepared by differential centrifugation, is resuspended and loaded onto a suitable continuous gradient, after centrifugation a number of visible bands will be found. If the gradient is fractionated and the fractions assayed for the presence of succinate dehydrogenase (SDH), it will be found to peak in one, or possibly two closely associated bands, thus determining the position of the mitochondria in the gradient. This does not necessarily mean that the mitochondria are in highly purified form; contaminants here can arise from two main causes: co-banding of particles of the same or very similar buoyant densities; or from the presence of particles still migrating towards their isopycnic position. The partially purified mitochondria pellet will contain many lysosomes, peroxysomes and smaller particles such as membrane-bound vesicles. Apart from possibly a few remaining nuclei, the mitochondria and lysosomes will be the largest particles present. When centrifuged on the continuous gradient, these large particles will sediment faster than the smaller ones and reach their isopycnic positions relatively quickly. Because they form the largest proportion of the sample, as they sediment they will carry some of the smaller particles with them. These particles may remain with the banded mitochondria and lysosomes or, if denser, sediment beyond them. The two organelle populations are also heterogeneous in size, so some will reach their banding density ahead of others. Hetero-geneity in density will compound the problem of cross-contamination with closely banding organelles.

Moreover, homogenization itself introduces variability [1]. It is very difficult to control the degree to which organelle fragmentation occurs from preparation to preparation. This is a particular problem with plasma membrane and vesicles.

We can now see the need for a thorough analysis of gradient fractions, especially following a new separation procedure. If we had confined our attention to the SDH profile of the gradient, we would not know of the presence of lysosomes (main contaminant) and other species present with the mitochondria. Of course, it may well be that these contaminants will have no effect upon subsequent work on the mitochondria, so that harvesting the fraction showing the peak SDH activity is sufficient. In that case, no further assays would be necessary, but it is clear that there is a need for full analysis of all fractions when new separations are being designed, so that the results of further investigations are not negated by the later realization of the presence of contaminants.

6.3 Removal of the gradient medium from fractionated material

Having fractionated the gradient, the treatment of the fractions will depend upon the subsequent studies to be carried out on the purified material and the gradient medium used. It may be necessary to remove all traces of the gradient medium from the fractions in order to carry out certain assays and some gradient media may interfere with the results of subsequent experiments, due to the media binding to the sample material, altering the conformation and function of molecules or having toxic effects, etc.

In many cases the medium can be removed by diluting the sample and washing the material several times by pelleting, but bear in mind that pelleting can damage delicate material and that loss of yield is almost inevitable. Other methods involve dialysis of the fractions against a suitable buffer, elutriation via a molecular sieve or ultrafiltration.

6.4. **Possible problems**

Almost everyone has experienced problems when seeking to repeat the protocols published by others and suffered the frustrations of obtaining completely different results from those published. In general, commercially prepared products to be used for specific purposes have been thoroughly tested for reproducibility of results, so, unless one receives a sample from a bad batch, there should be no problems, provided the protocol, which is usually detailed and precise, is followed exactly.

The isopycnic banding of particles will not be affected by using centrifugation speeds and times greater than those of the protocol because the particles will remain at their isopycnic positions. However, excessive centrifugation may damage delicate structures.

It is the methods which use differential pelleting or rate-zonal techniques that require more accurate centrifugation times and force fields. The distances that the particles travel to pellet or form zones will vary with changes in those parameters, thus affecting the purity of the fractions so obtained. Apart from any operator error in setting the centrifugation conditions, other errors may be due to the centrifuge used. Due to the wear and tear of usage, the true speed and temperature of the machine may not be registering accurately and so it may be spinning faster or slower than indicated. Such errors can easily be overcome by increasing or decreasing the indicated speed. Another error may occur in the timing device on the centrifuge and this can be easily checked against an accurate watch or clock.

We have discussed the dependency of the r.c.f. generated upon the speed and radius of rotation, and the importance of the pathlength in defining the distance a particle can move along the radius of rotation. It is obvious that as a particle moves along the centrifuge tube, its radius of rotation and thus the r.c.f. experienced, increases as it moves further from the axis of rotation. Therefore, when stipulating the r.c.f. for a particular purpose, it is important to state where this force is to be measured. The r.c.f. generated at a set r.p.m. in a particular rotor, is usually given by the manufacturer as g_{max}, the force at the maximum radius of rotation, while g_{av}, the force at the radius corresponding to midway along the pathlength of the rotor is quoted in the literature. The difference in the forces generated at these two points is considerable, especially with regard to rotors with long pathlengths. For example, a swinging-bucket rotor, spinning at 20 000 r.p.m., with a maximum pathlength of 20 cm, would generate about 90 000 g_{max}, but with a mid-pathlength of 15 cm would generate 67 000 g_{av}, showing the importance of indicating forces correctly in protocols. Giving centrifugation conditions in r.p.m. is only of value when the rotor used is also specified.

Attention to the above sources of error may save a considerable amount of frustration when setting up new centrifugation techniques or trying to repeat older ones.

Reference

1. Graham, J. (1975) in *New Techniques in Biophysics and Cell Biology* (R. Payne and B.J. Smith, eds). Wiley, New York, Vol. 2, p. 1.

7 Examples of the Use of Each Centrifugation Technique

The methods of applying the three centrifugation techniques which follow and the earlier discussion of the formation of density gradients, show that there are numerous possibilities open to experimentation when trying to separate biological particles. The nature of some material, such as blood cells or cultured cells, often allows a satisfactory purification of a particular species in a simple routine manner, as will be described later. In many cases, however, no such simple method has yet been developed and one has to devise methods by examining the density and the sedimentation rates of all the particles that may be present in a suspension. This may entail using all three of the techniques in order to obtain acceptable purity; differential pelleting to prepare a crude suspension of the particles; followed by either rate-zonal, isopycnic banding or both, for further purification. If rate-zonal separation produces a band of particles of similar size and sedimentation rates, but of different functions and buoyant densities, then isopycnic centrifugation of the particles from that band would be needed to resolve the two species.

In all separation methods, one has to consider the minimum degree of purity required that will allow further studies to be carried out with meaningful results. In general, the manipulation of material during the purification process inevitably leads to some damage and loss of material and, therefore, the higher the degree of purity, the lower the percentage recovery of the material. It would seem rather pointless to spend time and effort removing contaminants which would have no detrimental effect upon the results of work carried out in their presence. Thus, if one was examining the engulfing abilities of macrophages, it would be unnecessary to remove other cells from the suspension by complex methods, reducing recovery, if the macrophages were clearly recognizable and the presence of other species was known to have no effect upon the function of the macrophages. Moreover, protracted manipulation of the particles frequently leads to a deterioration of their biological activity. A decision to compromise between purity and recovery thus depends upon the subsequent studies to be undertaken.

7.1 **Differential pelleting**

A typical differential pelleting scheme is described here for obtaining partly purified suspensions of the several components of homogenized mouse liver. The exact compositions of the homogenizing solutions, buffers and gradient solutions depend upon the organelles to be purified and the subsequent investigations to be carried out using the purified material. For example, to maintain the integrity of nuclei, the presence of low concentrations (1–2 mM) of magnesium and/or calcium is required, while mitochondria exhibiting a high level of repiratory control, must be prepared in the presence of EDTA. The addition of other solutes, such as potassium, EGTA or mannitol is sometimes beneficial: the solutions described here will contain only Tris/EDTA in addition to the osmotic balancer (0.25 M sucrose). Homogenizing methods and media for the preparation of subcellular particles have been reviewed elsewhere [1].

This section is concerned with the centrifugation forces needed to pellet the various crude fractions, the specific conditions for each fraction are discussed in Section 9, 'Preparation of subcellular organelles'.

> *Homogenizing solution:*
> 0.25 M sucrose, 10 mM Tris/HCl (pH 7.6), 1.0 mM EDTA.
> This solution is iso-osmotic with the tissue but, as stated above, may require an alternative or addition to Tris/EDTA, depending upon the purpose of the separation.

All solutions and vessels should be kept *ice-cold*. At first sight, it might be considered beneficial to carry out the procedures at a temperature of 37°C, at which the cell functions optimally, but the degradative activity of the many enzymes that may be released during homogenization make this impractical.

A mouse is killed by cervical dislocation and the liver removed into ice-cold homogenizing solution (5–6 ml). The liver is cut into several large pieces and agitated in the medium with forceps to wash away as much blood as possible. The pieces are transferred to 3–4 ml of fresh medium and minced finely with scissors before tranferring them to the homogenizing vessel and adding homogenizing solution to a final volume of about 5 ml/g of liver. The tissue is homogenized using 8–10 strokes of the pestle and the homogenate strained through four layers of muslin to remove any undisrupted fragments and connective tissue. The filtrate is loaded into a 50 ml centrifuge tube and the nuclear fraction pelleted by centrifuging at 700 *g* for 10 min in a fixed-angle rotor, such as an 8 × 50 ml rotor.

The post-nuclear supernatant is decanted into a clean 50 ml centrifuge tube and centrifuged at 10 000 *g* for 10 min to pellet the mitochondria and lysosomes. The supernatant is again decanted into a fresh tube and centrifuged, this time at 100 000 *g* for 1 h to pellet the microsomal fraction.

The supernatant will now contain the soluble fraction of the homogenate, consisting of cytoplasm and soluble components released from any broken organelles.

This very crude differential centrifugation exercise will provide very heterogeneous pellets. Additional centrifugation steps may improve the resolution (e.g. a 3000 g for 10 min step to pellet heavy mitochondria — but not lysosomes). Resuspending the pellets and recentrifuging will wash away some of the contaminating material but this leads to loss of material with each washing cycle.

7.2 **Rate-zonal centrifugation**

The rate-zonal separation of particles requires a supporting medium, generally a continuous density gradient of sucrose, to be prepared with a density range such that all the particles in the sample will sediment through it, given sufficient time and speed of centrifugation. The theory is that all the particles will begin to sediment from approximately the same area (thus the sample must be applied to the top of the gradient as a narrow band), at a rate proportional to their size and density. As they progress through the gradient, the particles' rate of sedimentation will be slowed by the increasing density and viscosity of the gradient, but increased by the higher r.c.f. as they move further along the axis of rotation (see *equation 1*, Section 1.2). Together, this results in an almost constant sedimentation rate through the gradient for each type of particle. The faster sedimenting particles will draw ahead of the slower, so forming bands or zones of particles of similar size and density. The rate at which particles sediment is related to a unit of sedimentation called a Svedberg unit, 'S'. The derivation of this unit is not important here, but see Spragg [2] for a detailed description. It is only necessary to understand that the higher the S-value, the faster the sedimentation rate. It can be visualized that after rate-zonal centrifugation, the particles with higher S-values will have travelled further along the gradient than those with lower values. From intensive studies of these factors, some common biological particles have been assigned S-values which make their separation by rate-zonal techniques simple to plan. For example, ribosomes are made up of a number of RNA molecules, in the case of bacterial ribosomes the whole ribosome is assigned a value of 70S and the various RNAs 5S, 16S and 23S. It follows then, that after rate-zonal centrifugation of the molecules, the 23S unit will be near the bottom of the gradient, the 5S near the top and the 16S somewhere in between. For a given gradient profile, the distance travelled by a particle of a given S-value will depend upon the density and viscosity of the gradient medium, the centrifugation time and the centrifugal field applied. The density and viscosity of the medium will change as the temperature changes so it is important to keep the temperature constant during the centrifugation.

There are computer programs available that have been devised to enable one to choose the correct parameters for a rate-zonal run, using sucrose as the supporting medium. The programs require the operator to feed in details of the rotor to be used, the gradient volume, sample volume, the number and volume of the fractions into which the gradient will be fractionated, and the time, speed and temperature of centrifugation.

Some programs may request the concentration of the medium at the top and bottom of the gradient, from which the computer calculates the concentration of medium in each fraction of a perfectly linear gradient. Other programs require the operator to input these concentrations. In either case, having information of the gradient profile and the speed, time and temperature of centrifugation proposed, it is able to calculate the S-values of particles which would be in each gradient fraction and display them on the screen.

We are thus able to see which of the fractions, of the theoretical gradient specified, would contain the particles in which we are interested, say, for example, the 16S and 23S units of ribosomal RNA. If it can be seen that they are not sufficiently separated, the conditions can be adjusted by changing various parameters input to the program until a satisfactory result is shown (see *Table 7.1,* A and B).

Using this as a guide, we set up the conditions indicated and carry out the centrifugation. Various factors may combine to cause the gradient to deviate to some extent from the ideal conditions specified. The gradient is unlikely to be exactly linear and small variations during centrifugation will alter the expected sedimentation rates. After centrifugation, it is therefore necessary to fractionate the gradient into the specified number of fractions, measure the concentration of each fraction, usually by refractive index,

TABLE 7.1: *S-values for computer-simulated gradients A and B, and for an actual gradient, prepared by reference to the simulations, C,D (see text for details).*

	A		B		C		D		
No.	% (w/w)	S-value	% (w/w)	S-value	% (w/w)	S-value	No.	% (w/w)	S-value
1	10.5	2.24	10.5	2.79	10.8	2.82	7	13.25	11.53
2	11.5	5.12	10.5	6.38	11.4	6.40	8	13.75	13.20
3	12.5	7.89	12.5	9.84	12.2	9.81	9	14.25	14.84
4	13.5	10.58	13.5	13.19	13.0	13.10	10	14.75	16.46
5	14.5	13.20	14.5	16.45	14.2	16.32	11	15.25	18.07
6	15.5	15.76	15.5	19.65	15.6	19.54	12	15.75	19.66
7	16.5	18.29	16.5	22.80	16.7	22.71	13	16.25	21.24
8	17.5	20.78	17.5	25.92	18.0	25.90	14	16.75	22.81
9	18.5	23.26	18.5	29.01	19.3	29.10	15	17.25	24.37
10	19.5	25.79	19.5	32.00	20.2	31.32	16	17.75	25.92

and return to the computer program.

The information is input as before, but now, with accurate knowledge of the gradient profile from the measurement of the actual gradient fractions, the ideal linear gradient specified is replaced with the information on the actual gradient-profile measured. This allows the computer to calculate a new set of S-values for each fraction and, hopefully, if the gradient has been made efficiently, not deviating too much from the ideal gradient initially specified. The list displayed then identifies the fractions which contain the particles of interest.

Table 7.1 shows the S-values for three gradients, A, B and C. All are 10 ml in volume, 10–20% (w/w) sucrose and all are fractionated into 10 × 1 ml fractions. A and B are computer-simulated, the values for an ideal linear gradient having been input. C is an actual preformed continuous gradient, of 10–20% (w/w) sucrose, fractionated into 10 × 1 ml fractions and the concentration of each measured after centrifugation. Gradient A is a simulation in which the centrifugation conditions were: 25 h at 25 000 r.p.m. at a temperature of 4°C. It can be seen that the 16S and 23S particles are in fractions 6 and 7, and 8 and 9, respectively. Gradient B is identical except for time, being centrifuged for 20 h and resulting in a better resolution of the particles — 16S being in fractions 4 and 5 and 23S in fractions 7 and 8. The experimental gradient, C, was set up on the same parameters as B and the measured concentrations of the fractions entered into the computer. In this experiment, gradient C was very close to the theoretical gradient, B, and therefore the particles were found to be in the same fractions as the theory predicted. Taking smaller gradient fractions isolates the particles of interest within a smaller volume and within a smaller range of S-values. Column D shows the results of fractionating gradient C into 20 × 0.5 ml fractions. For simplicity, only fractions 7–16 are shown, together with the S-values for each fraction. The 16S and 23S particles are now in fractions 9 and 10, and 14 and 15, and contained in 2 × 0.5 ml volumes instead of 2 × 1 ml.

The temperature is a very important factor during rate-zonal centrifugation so the centrifuge, rotor and the gradients should all be pre-cooled to the correct temperature before the rate-zonal run begins. It is also advisable to run a 'spare' gradient at the same time, so that the temperature of the gradient can be checked immediately the centrifuge stops, using an accurate, pre-cooled, thermometer with the spare gradient. Any deviation from the chosen temperature can then be entered into the computer. Setting up different conditions on the computer can be done in minutes and allows the determination of the correct conditions for a rate-zonal separation.

As well as determining the positions of particles of known S-values, the programs can also be used to obtain some idea of the S-values of particles from their positions in a gradient. In this case, a gradient is set up and the sample centrifuged as described. The gradient is fractionated and the concentration of the gradient medium in each fraction determined. The position of the particles of interest are then found by assaying each fraction. By entering the gradient profile and the centrifugation conditions into the computer program, the S-value of each fraction is displayed. Knowing the fractions which contain the particles of interest from the assay procedures, a range of S-values for the particles can be computed. Referring to *Table 7.1C*, if the assay indicates that fraction number 6 contains the particles, we know their S-value is between 16.32 and 22.71. By varying the density range of gradients and/or centrifugation conditions, we can use the computer to simulate gradients which will cover S-values from, say, 25S to 38S and, by fractionating into smaller fraction volumes, for example 0.2 ml instead of 1 ml, we can make further centrifuge runs to obtain a smaller range of the S-values which contain our particles. While rate-zonal banding as described will give some idea of the S-values sought, an analytical centrifuge is required for very accurate determination of these values. A brief description of analytical and other specialized centrifugation equipment is found in Appendix A.

7.3 Isopycnic banding

Isopycnic banding refers to banding particles in the area of a gradient at which the density of the particles equals that of the surrounding medium and further migration ceases. A continuous gradient is used, either self-forming or preformed, depending upon the particles to be separated. The gradient has to be sufficiently dense at the bottom of the tube to prevent the material of interest pelleting, and not so dense at the top of the tube as to prevent the particles entering the gradient, or, in the case of samples bottom-loaded or distributed throughout the gradient initially, to cause them to float to the top.

Manipulation of the gradient profiles allows the resolution of bands of material to be enhanced. For example, two species with very similar buoyant densities will overlap each other in a steep gradient, but the two will be more clearly resolved (more clearly separated) on a shallow gradient (see *Figure 5.1*). It is best if the density range of the gradient is such that the band of material of main interest is formed near the middle of the gradient. Once the particles have reached their isopycnic position, no further migration takes place, but it should be kept in mind that the various particles will take different times to reach their particular isopycnic positions, depending upon their particular sedimentation rates.

In practice, this means that if a band of material is formed at its isopycnic position in a relatively short time, it may later be contaminated with more slowly sedimenting particles which may be denser, but are still moving to their own isopycnic positions, and therefore a more purified fraction might be obtained using shorter centrifugation times. This demonstrates the necessity, especially when carrying out new separations, to fractionate and analyse the whole gradient, and to assay for species other than those of main interest. This will indicate the positions of potentially contaminating material and the manner in which they are sedimenting. Such information can then be used in order to enhance the purity of fractions.

A useful method of examining the migration of particles during centrifugation is to prepare sets of identical density gradients and load one sample on top and the other at the bottom of pairs of gradients. If each pair are centrifuged for different time periods, but at the same rotor speed, the gradients can be fractionated and analysed to examine the rates at which each type of particle is moving, giving useful information with regard to designing a gradient for a particular purification.

If a high-speed or ultracentrifuge is used, it is necessary to consider balance of the tubes. Therefore prepare identical continuous gradients, leaving sufficient room in the centrifuge tubes so that, for each pair, one tube has a bottom-loaded sample in dense medium and the other tube an equivalent volume of dense medium. The top-loaded tube is balanced by the other tube having an equivalent volume of the medium in which the top-loaded sample is suspended. This will ensure the tubes are correctly balanced for high-speed and ultra-speed runs. If very precise calculations were to be made, it would be necessary to consider the differences of the r.c.f. experienced by the the bottom-loaded and top-loaded samples, but that is not a matter for us to consider in this book.

7.3.1 *Isopycnic banding of mitochondria*

Mitochondria from mouse liver are commonly prepared from the mitochondrial pellet of mouse-liver homogenate as described in Section 7.1. The resuspended pellet is centrifuged on a 1–2 M continuous sucrose gradient for 2 h at about 75 000 g. Two bands of mitochondria are then present, one the intact organelles, the other broken or damaged mitochondria. On sucrose gradients, mitochondria band at a buoyant density of about 1.2 g/ml due to the very hyper-osmotic conditions of the gradient, while on gradients of Nycodenz or metrizamide where conditions are near iso-osmotic, they band at close to 1.15 g/ml. On continuous Percoll gradients they have been banded at buoyant densities of 1.11–1.13 g/ml. Due to their lower viscosities, shorter centrifuge times and lower r.c.fs are required on gradients of Percoll and the iodinated media.

If we consider the sedimentation rates and buoyant densities of a top-loaded sample, centrifuged on a 10–40% (w/v) Nycodenz gradient, it will demonstrate how the purity of fractions are affected by the period of centrifugation. The sample material quoted above will contain — in addition to the mitochondria — ribosomes, nuclei, lysosomes, peroxisomes and membrane vesicles. We can make an educated guess at what occurs when centrifugation commences. Any nuclei in the sample will sediment faster than the mitochondria and, as their buoyant density is around 1.23 g/ml, they will stay ahead. The mitochondria and lysosomes, although both quite heterogeneous in their sizes, will be the next fastest sedimenting particles, and will band at a buoyant density of about 1.15 g/ml. Peroxisomes, with buoyant densities of 1.23–1.25 g/ml and ribosomes and their subunits, 1.2–1.35 g/ml, will be following the mitochondria and, given sufficient time, will pass through the mitochondrial band to find their own isopycnic density.

It is clear that, if the sample is loaded as a fairly narrow band, centrifugation starts with a rate-zonal separation, but the initial widening of the gap between the bands is stopped when the faster sedimenting, but less dense, particles reach their isopycnic position and remain there, allowing the denser, but slower moving, particles to overtake them. On true rate-zonal gradients the density is such that all the particles for separation can keep sedimenting and eventually pellet.

At the centrifugation time and forces employed, the soluble components and very small particles will barely enter the gradient from the loading area. It is possible, then, to have the mitochondrial fraction banded at its isopycnic position some time before the ribosomes and peroxisomes, etc. arrive at that point and the centrifugation time and speed can be adjusted to achieve this result.

References

1. Graham, J. (1975) in *New Techniques in Biophysics and Cell Biology* (R. Payne and B.J. Smith, eds). Wiley, New York, Vol. 2, p. 1.

2. Spragg, S.P. (1978) in *Centrifugal Separations in Molecular and Cell Biology* (G.D. Birnie and D. Rickwood, eds). Butterworth, London, p. 7.

8 *Preparation of Intact Cell Populations*

In order to separate a mixture of different types of cells into purified fractions of each type, or, more usually, a purified fraction of one particular type, a single-cell suspension of the mixed cells is required. In samples such as peripheral blood, amniotic fluid or cultured cells growing in suspension, the cells are collected together with their immediate environment, serum in the case of peripheral blood, the amniotic fluid itself, and, in the case of cultured cells, the medium in which they have been growing and dividing, often for many generations. These cells, then, are not subjected to any harsh treatment during collection which might damage or alter their functions, at least not immediately. However, cell suspensions which are prepared by dissociating whole tissues or organs by enzymatic digestion and cultured cells growing in monolayers which have to be lifted by, for example, trypsin treatment, may suffer damage and changes in form and function. Methods for the preparation of single-cell suspensions from tissues and monolayer cultures are beyond the scope of this book but have been fully described elsewhere [1,2].

The importance of the influence of the osmotic environment upon the buoyant density of cells and other membrane-bound particles has already been examined in the discussion of the properties of gradient media. It is therefore clear that the choice of an appropriate gradient medium is paramount when cells are involved. In the following pages we will examine methods for the fractionation of mixed cell populations into their different species, using continuous and discontinuous gradients and both isotonic and hypertonic conditions.

8.1. **Purification of blood cells**

Blood is a readily available example of a single-cell suspension of mixed cell types and so provides a good starting point for looking at methods of preparing purified fractions of intact cells. The extensive uses of blood to assist in the diagnosis of health problems can be deduced from the numbers of patients lining up for blood tests at any hospital. The haemotology and biochemistry departments of the hospitals need to be able to carry out the tests called for as rapidly and accurately as possible. The routine methods that are used in the hospital departments must, therefore, be as simple to carry out as possible and give consistently reproducible results. If a pure preparation of a single component of blood is required for examination, e.g. the platelet population, a rapid and reliable method, which does not affect the function or morphology of the platelets, is needed. More complex methods of purification may be used for research purposes and in the search for new methods, but for routine diagnostic work, a simple, efficient protocol is most necessary. To this end, a great amount of effort has gone into preparing such methods.

Peripheral blood contains a number of types of cell differing in both form and function, divided into two main classes, red cells (erythrocytes) and white cells (leucocytes). The leucocytes are divided into five sub-populations, lymphocytes and monocytes (mononuclear cells), and neutrophils, basophils and eosinophils (the granulocytes). The peripheral blood also contains platelets, small non-nucleated cells with functions related to blood-clotting. There are about 700 erythrocytes per leucocyte in human blood, but the relative proportions of the different populations of cells vary greatly among individuals. Moreover, the relative proportions found in a sample also varies with the state of health of the individual.

Under isotonic conditions, the mononuclear cells have buoyant densities of about 1.06 g/ml with very little difference between the lymphocytes and monocytes; basophils about 1.07 g/ml; neutrophils 1.085 g/ml; and eosinophils 1.095 g/ml. Erythrocytes have buoyant densities ranging from about 1.09 to 1.11 g/ml. It can be seen that the mononuclear cells and the basophils have fairly close buoyant densities, while the neutrophils and eosinophils (polymorphonuclear cells) together with the erythrocytes also have closely similar densities, but are well separated from those of the mononuclear cells. These characteristics offer a way to separate the mononuclear cells from the polymorphonuclear cells and erythrocytes.

8.1.1 *Leucocyte-rich plasma*

In some of the techniques that follow, a leucocyte-rich plasma (LRP) is required. This refers to the supernatant which remains after blood has been left to stand long enough to allow most of the erythrocytes to sediment, leaving a cloudy, straw-colored supernatant above the pellet. In order to aggregate the erythrocytes and thus speed their sedimentation, a high molecular weight polysaccharide, such as dextran, can be added. The dextran solution is prepared by dissolving 6 g of dextran (mol.wt 500 000) in 100 ml of 0.9% (w/v) NaCl solution.

If 1 part of the aggregating solution is added to 10 parts of blood, mixed and allowed to stand at room-temperature, the erythrocytes will aggregate and start to sediment. The time taken for the erythrocytes to sediment until the pellet occupies about half or less of the total volume, varies with different donors, 15–45 min is the usual time range. The straw-colored supernatant above the pellet of red cells is the LRP, which may be harvested by drawing it off with a Pasteur pipette and including the diffuse red-tinged area at the top of the pellet. Alternatively, the red cells can be removed using a syringe fitted with a metal cannula, which is introduced carefully to the bottom of the tube and the erythrocyte pellet withdrawn. The LRP will contain between 60 and 80% of the total leucocytes of the sample, most of the platelets and many erythrocytes, but the erythrocytes will now be greatly depleted, about twenty or so per leucocyte. The leucocytes and erythrocytes can be pelleted, if required, by centrifuging at about 150 g for 5 min. The supernatant, a platelet-rich plasma, can then be centrifuged at 2000 g for 15 min to pellet the platelets, leaving a cell-free autologous plasma for resuspending the leucocytes.

8.1.2 *Preparation of mononuclear cells*
The differences in buoyant density allow the mononuclear cells, lymphocytes, monocytes and basophils, to be separated from the polymorphonuclear granulocytes and erythrocytes by a simple, one-step procedure, first developed by Bøyum [3]. A density barrier is prepared which allows the erythrocytes and granulocytes to pass through, but retains the mononuclear cells at the sample/medium interface. The separating medium developed by Bøyum, which has been widely used as the standard method for the preparation of mononuclear cells, is a solution of 9.6% (w/v) sodium metrizoate, and 5.6% (w/v) Ficoll (a high molecular weight synthetic polysaccharide) which provides a solution with a density of 1.077 g/ml and with an osmolality of 300 ± 5 mOsm.

Blood is collected into a tube containing a suitable anti-coagulant, heparin or EDTA are the most commonly used and give the best results in this separation. However, different anti-coagulants have different effects upon the cells and it may be necessary, in some cases, to use an alternative anti-coagulant, e.g. citrate. The blood sample should be diluted by addition of an equal volume of physiological saline and carefully overlayered onto the separating medium in a centrifuge tube in the ratio of 1 volume of medium to 2 volumes of the diluted blood. The dilution of the blood reduces the concentration of the cells so reducing the effects of artifactual banding, which can occur when large concentrations of particles are migrating and taking unrelated particles with them, e.g. erythrocytes being trapped in the mononuclear layer, or mononuclear cells being pulled down with the erythrocytes. When the gradient is centrifuged at 800 g for 15 min, in a swing-out rotor, the red cells will pellet and a lighter, buff-colored layer will be seen on top of the pellet. There will be a prominent, whitish-colored band at the interface of the sample and medium, which will contain the mononuclear cells and platelets. A light-colored layer on top of the pellet will contain the polymorphonuclear granulocytes.

The widespread use of Bøyum's procedure has meant that commercially prepared separating solutions under several different trade names, such as Lymphoprep or Ficoll-Hypaque are available, although in some cases sodium diatrozoate is substituted for metrizoate and a different polysaccharide may be used. In essence the results obtained are very similar.

8.1.3 *Removing monocytes from the mononuclear cell fraction*
The monocytes are phagocytic cells and this property can be used to remove them from the mononuclear cell-band. If the blood sample is centrifuged at 150 *g* for 5 min, most of the red cells will pellet leaving a cloudy, LRP as supernatant and a buffy-colored coat of leucocytes lying on top of the pellet. This is quicker than the dextran sedimentation method described earlier. Collect the supernatant together with the buffy coat (disturbing the erythrocytes as little as possible), and incubate in a shaking water bath at 37°C for 30 min in the presence of colloidal iron.

The monocytes will ingest the iron particles, thus increasing their buoyant density. If the LRP is now layered on top of the separating medium and centrifuged as before, the monocytes will sediment with the erythrocytes and granulocytes leaving a band of lymphocytes and platelets at the interface.

8.1.4 *Isolation of monocytes*
The isolation of pure monocytes requires different conditions to be met. Under isotonic conditions, monocytes have a slightly lower buoyant density than lymphocytes, but the difference is not sufficient to allow them to be separated across a single density interface as described above. On a continuous density gradient the mononuclear cells form a rather broad band with the top of the band enriched in monocytes and the bottom depleted. As the monocyte population is only 6–8% of the total leucocytes, they cannot be harvested in a pure form from the band.

The way in which cells tend to lose water and shrink in hypertonic conditions has been studied in a systematic manner by Bøyum [4]. He found that lymphocytes are more sensitive to osmotic changes than monocytes and that, under hyper-osmotic conditions, the change in buoyant density due to water loss was greater in lymphocytes than in monocytes. These findings allowed Bøyum to develop a medium for the purification of monocytes using a solution of Nycodenz together with sodium chloride as an osmotic balancer. The solution has a slightly lower density than the Ficoll-sodium metrizoate solution, 1.068 g/ml, and a higher osmolality, 330 mOsm. When LRP is layered on top of this solution and centrifuged at 600 *g* for 15 min, the erythrocytes and granulocytes pellet as before. The decreased density of the medium allows the mononuclear cells to migrate into the medium with the lymphocytes sedimenting at a slightly faster rate than the monocytes due to their differential increase in density over the monocytes. After centrifugation we will therefore have a situation where the concentration of lymphocytes is high near the red cell pellet and decreasing towards the top of the tube, with the reverse true for the monocyte concentration. Harvesting cells from the top of the medium downwards, the purity of the monocytes will be highest in the top fractions with lymphocyte contamination increasing in the lower fractions. This gives a choice of high purity with relatively low recovery, or higher recovery of the monocytes with decreasing purity. The earlier discussion of the principles of rate-zonal separations indicates that better resolution of the two species will be obtained if they are loaded as a narrow band and this can be done by pelleting the cells of the LRP by centrifugation at 250 g for 10 min and resuspending the pellet in a smaller volume before layering on top of the separation medium.

8.1.5 *Isolation of polymorphonuclear cells*
The polymorphonuclear cells (PMNs) in human blood consist of neutrophils and eosinophils, the neutrophils comprising 40–75% of the total leucocytes and eosinophils 1–6%.

(a) A rapid, single-step method for the isolation of PMNs. Simplified, routine methods have been devised for the one-step preparation of PMNs from whole blood, again utilizing the effects of changes of osmolality upon the buoyant density of cells.

The buoyant density of erythrocytes ranges from about 1.09–1.11 g/ml under isotonic conditions with neutrophils about 1.085–1.09 g/ml and eosinophils 1.09–1.1 g/ml. These closely similar buoyant densities make them very difficult to separate consistently using a single-step solution, as is possible with the mononuclear fraction. However, by utilizing the knowledge that different cells lose water at different rates under hyper-osmotic conditions [4], experimental adjustments of density and osmolality have allowed solutions to be devised which permit polysaccharide-aggregated erythrocytes to pellet and PMNs to increase their buoyant density sufficiently to allow them to migrate slowly into the medium while the mononuclear cells remain at the interface.

Commercially prepared solutions, such as Polymorphprep® (Nycomed) or Mono-Poly-Resolving Medium (Flow Laboratories) are available. The solutions have similar properties: density 1.112–1.114 g/ml and osmolalities 450–500 mOsm.

Human blood is collected using heparin or EDTA as anti-coagulant and the undiluted blood is layered on top of the medium (5 ml blood on 5 ml medium) in a centrifuge tube. After centrifugation at 450 g for 35–40 min at a temperature of 18–23°C, the red cells are pelleted, the mononuclear cells banded at the sample/medium interface and the PMNs banded about 5 mm below the mononuclear band. The bands can be harvested using a Pasteur pipette. The harvested PMN band should be diluted by addition of an equal volume of 0.45% NaCl solution (or medium of equivalent osmolality) in order to restore normal osmolality, remember that the separating medium has an osmolality of 450 mOsm. Further dilution of the density by addition of 2 volumes of normal saline allows the cells to be pelleted by centrifugation at 250 g for 10 min.

This separation relies on the dextran-aggregated erythrocytes losing sufficient water and increasing their density to a degree that allows them to sediment relatively quickly through the dense solution, while the PMNs also lose water but migrate more slowly than the erythrocytes, forming a band between the mononuclear cells — which are unable to enter the medium — and the pelleted erythrocytes. This situation holds at a temperature of 18–23°C, at higher temperatures the density and viscosity of the solution decrease sufficiently to allow the PMNs to sediment more quickly and the band becomes very diffuse with the PMNs distributed through the medium or all lost to the pellet. Therefore, when using a centrifuge without a temperature control, it is important in warm room conditions to check that the temperature is within the range given.

In some cases it may be advantageous to isolate neutrophils which have not been exposed to high molecular weight polysaccharides. There have been reports of exposure to such polysaccharides inhibiting a number of neutrophil functions: impaired ability to migrate under agarose [5,6]; stimulation of the metabolism of the cells [7]; and partial irreversible binding to the cell and alteration of the surface charge [8]. The isolation of PMNs using Ficoll or Percoll is reported to cause the cells to lose their ability to adhere and spread normally on glass or plastic surfaces [9]. While these effects need to be considered when they are likely to interfere with subsequent investigations, many methods which entail exposure of the cells to polysaccharides are used successfully.

To isolate PMNs, when either the polysaccharide methods are not advisable, or when one wishes to compare the results with and without polysaccharides, the following procedure [10] may be used.

(b) Purification of PMNs on continuous density gradients. If one attempts to separate whole blood on a continuous, isotonic density gradient, the band of PMNs is completely masked by the sheer numbers of erythrocytes present. Therefore, it is necessary to remove the majority of erythrocytes by preparing a LRP. In this case, in order to avoid exposure to polysaccharides, the blood may either be allowed to stand at room temperature until sedimentation has occurred without an erythrocyte-aggregating medium being added (this will take longer than a preparation with dextran), or the whole blood may be centrifuged at about 200 g for

5 min and the resulting supernatant and buffy coat removed with a syringe and cannula or Pasteur pipette.

An isotonic, continuous gradient of 5–20% (w/v) Nycodenz is prepared (see Section 5 for details) and the erythrocyte-depleted cells loaded on top and centrifuged for 20 min at 1000 g at about 20°C. Mononuclear cells are banded near the top of the gradient and the erythrocytes and PMNs form two bands towards the bottom of the tube, the neutrophils above the erythrocytes. Depending upon the individual blood samples and the exact conditions of centrifugation, the two bands may be clearly separated or overlapping. The number of erythrocytes present in the initial LRP will of course influence the resolution of the separation. It is therefore necessary to take particular care, when collecting the LRP and buffy coat, to reduce the numbers of erythrocytes taken from the top of the erythrocyte pellet.

8.1.6 *Isolation of platelets*

Human blood platelets are used in the study of blood-clotting and a number of ailments. A simple, routine method of obtaining populations of platelets, free of other cells and in a functionally intact state has been developed [11] and is described here. Platelets are the smallest cells of peripheral blood but are quite heterogenous in size and density. Due to their small size in relation to the other blood cells, their sedimentation rate is relatively slow. This difference in sedimentation rates is exploited in the following protocol.

A solution of 12% (w/v) Nycodenz, 5.6% (w/v) NaCl and 5 mM tricine/NaOH (pH 7.4) has a density of 1.063 g/ml and an osmolality of 315 mOsm. This solution allows all the blood cells to sediment through it, given sufficient centrifugation time and force.

If undiluted whole blood, collected using citrate as the anti-coagulant, is carefully layered on top of the medium in a centrifuge tube, 5 ml blood on 5 ml medium, and centrifuged at 350 g for 15 min at 20°C, the distribution of the cells will be as shown in *Figure 8.1*. All erythrocytes and granulocytes are pelleted, as are most of the mononuclear cells. The platelets are in a wide band, concentrated near the sample/medium interface and becoming less concentrated towards the pellet. As shown, 70% of the total platelet population can be harvested between the interface and a point about 10–15 mm into the medium, virtually free of any other contaminating cells, whereas harvesting to within about 5 mm of the pellet will recover more than 95% of the platelet population. However, the closer to the pellet, the greater the contamination by other cells.

This simple one-step method allows recovery of the platelets in a fully functional state for use in further studies.

8.1.7 *Separation of blood cells on Percoll gradients*

The use of Percoll gradients for the purification of the various species of

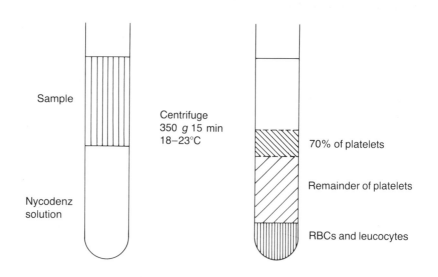

Sample

Centrifuge
350 *g* 15 min
18–23°C

70% of platelets

Remainder of platelets

Nycodenz
solution

RBCs and leucocytes

FIGURE 8.1: *Purification of platelets. A volume of whole, citrated blood is loaded onto an equal volume of the separating solution and centrifuged in a swing-out rotor for 15 min at 350 g. The distribution of platelets after centrifugation is shown.*

blood cells has been widely reported (full up-to-date details can be found in the *Methodology and Applications Booklets*, obtainable from Pharmacia-LKB), but no Percoll solutions for the routine isolation of particular cell types are currently available. Solutions for the different applications must therefore be prepared by dilution of the sterile Percoll solution supplied, as described earlier (see Section 5.4, 'Formation of iso-osmotic gradients'). Using preformed, iso-osmotic Percoll gradients, with a density range of 1.03–1.15 g/ml, the banding densities of the different cell types of human blood are very close to those found on iso-osmotic Nycodenz gradients of the same density range.

On discontinuous Percoll gradients, enhanced fractions of T and B lymphocytes [12], purified Natural Killer (NK) lymphocytes [13] and, on a 5-step gradient, the separation of lymphocytes and monocytes [14] have been described. Erythrocytes have been subfractionated on both discontinuous [15] and continuous [16] Percoll gradients.

Resolution of T and B lymphocytes has been achieved on discontinuous Percoll gradients [12]. Iso-osmotic aliquots of Percoll are prepared at densities of 1.052, 1.063, 1.075, 1.085 and 1.122 g/ml. The aliquots are layered over a sample of mononuclear cells to form a discontinuous, bottom-loaded gradient. After centrifugation at 2000 *g* for 10 min B cells and macrophages were found at the 1.052/1.063 g/ml interface and the T cells in two bands, at the 1.063/1.075 and 1.075/1.085 g/ml interfaces.

8.2 Cells from other body fluids

Single-cell suspensions occur naturally in body fluids other than blood: spermatozoa in different stages of development in the testes and in ejaculates; cells in amniotic fluid; cells obtained from lavaging body cavities such as the peritoneal cavity or lungs. The separation of these cells into their discrete species may be possible on density gradients by similar techniques to those used for blood cells. Although the cells are obtained as single-cell suspensions, dead, damaged and exfoliated cells are likely to be present. It is usual to wash and concentrate the cells by pelleting and resuspending them and this operation can lead to further damage. Damage can be minimized by sedimenting the cells onto a cushion of gradient medium as described in the following procedure.

Cells obtained by lavage of the lungs are collected in a fairly large volume (about 120 ml) and it is necessary to concentrate the cells before attempting to separate the different cell types present in the lavage. When this is done by pelleting and resuspending the pellet in a smaller volume, it is found that the percentage of viable cells is reduced, as judged by trypan blue exclusion. When the cell samples are centrifuged on either continuous or discontinuous, isotonic Nycodenz gradients, strings of aggregated material appear along the length of the gradient, disrupting the banded cells.

The damage is minimized by sedimenting the cells onto a 2 ml cushion of 28% (w/v) Nycodenz solution and the cells banded at the interface formed. By withdrawing the supernatant to within 2 ml of the cell band, the banded cells can be gently mixed with the Nycodenz and remaining supernatant to give a cell suspension in 14% (w/v) Nycodenz. When this sample is loaded into the centre portion of a discontinuous, iso-osmotic Nycodenz gradient and centrifuged for 30 min at 1000 g, cells band at each interface and no strings of aggregates are present (T. Ford, unpublished data).

The above indicates that when dealing with cells from body fluids, problems of damage and aggregation can be minimized by not allowing the cells to be subjected to severe compressive forces during washing and concentrating steps.

8.3 Separation of cells from organs and tissues

The separation methods described for cells from blood and other body fluids demonstrate some of the difficulties encountered in trying to purify cells from mixed populations and the techniques that may be used to overcome them. The various combinations of density and osmolality that have been used to enhance difficult separations can equally be applied to mixed cell populations from other sources. However, the preparation of single-cell suspensions from whole organs or tissue samples can present further problems. The enzymes used to break down the intercellular matrices of, for example, collagen fibres, can have adverse effects upon the integrity of the cells resulting in many dead cells being present and their contents free in the suspension. The cellular debris, especially long strands of DNA, can have the effect of causing the cells to bind together in clumps during centrifugation and thereby prevent efficient separation of the different species. This is very noticeable in the gradients, especially continuous gradients, by the presence of long, cork-screwing strings of material throughout the gradient. While it is beyond the scope of this book to provide detailed methods for the dissociation of organs and tissues, we can briefly look at some of the problems and discuss methods of overcoming them. (For detailed information on tissue and organ dissociation, see references 1 and 2.)

8.3.1 *Whole organ perfusion*
Liver consists of a number of cell types of different sizes and functions, divided into parenchymal and non-parenchymal cells. The parenchymal cells, or hepatocytes, form by far the largest proportion, about 65% of the total, and are also the largest cells of the liver, approximately 10 times the volume of the average non-parenchymal cell. The non-parenchymal cells are sub-divided into Kupffer cells (macrophages), stellate cells (fat-storing cells) and endothelial cells: they represent 10%, 3–4% and about 20–22% of the total liver cells, respectively. In order to study the function of each type and observe the passage of metabolites, etc. passing through the liver, it is necessary to be able to prepare a single-cell suspension and separate the cells into their different classes.

A single-cell suspension of rat-liver cells can be prepared by perfusion of the liver with an enzyme solution to break down the intercellular matrix and free the cells. A full description of one of the perfusion methods can be found elsewhere [17], but the method will briefly be discussed here.

Desmosomes, which form the cell–cell junctions are calcium dependent and, in the absence of calcium, the junctions break down. Intercellular matrices of fibres, such as collagen, can be broken down by enzyme digestion and in combination, these processes free the individual cells in the encapsulating membrane of the liver. The liver is first perfused *in situ* with a calcium-free solution via the portal vein, to flush away most of the blood and start the breakdown of the intercellular junctions. After about 10–15 min of this initial perfusion, the enzyme solution, in this case, collagenase, is introduced to the perfusion system. The action of collagenase is calcium-dependent and therefore calcium is present in the solution. After about a further 10 min perfusion, it is possible to see the disintegration of the liver within its capsule. The liver is then removed to a dish containing incubation medium, and the capsule split and agitated to release the cells into the medium. The fragments of capsule and any remaining large aggregates are removed, leaving a single-cell suspension of liver cells.

During the enzyme perfusion step a large proportion of the cell-surface receptors are lost, but it has been found that incubation of the suspension at 37°C for 45 min restores a large proportion of such receptors.[18]. The separation of the cells into their respective classes, or the purification of a specific cell-type, can now be considered.

The very large size of the parenchymal cells, as compared with the non-parenchymal, immediately suggests that differential centrifugation should provide a method of purifying this fraction of the liver. The great numbers of these cells also means that washing the pellet several times, with the subsequent loss of material entailed, should still be able to provide a large number of hepatocytes in a very pure form. Centrifugation of the cell suspension at low speed (about 50 g) for 15–20 sec is all that is required. The low-speed centrifugation does not sediment the dead hepatocytes and so many of them will be retained in the supernatants, which are, of course, enriched with non-parenchymal cells. As expected, the recovery of parenchymal cells in pure form is a relatively simple operation, but the non-parenchymal cells present some problems.

Because the dead and damaged hepatocytes do not pellet during the low-speed centrifugation steps, the pooled supernatants, in which the majority of non-parenchymal cells remain, will also selectively retain the dead parenchymal cells. While repeated centrifugation at 600 g for 4 min at 4°C will result in a fairly pure preparation of the non-parenchymal cells, the recovery is low, only about 25% of the initial cells in the pooled supernatants are recovered and there is a selective loss of the Kupffer cells (up to about 50%).

To increase the yield of non-parenchymal cells, agents which can destroy the parenchymal cells have been used. Pronase [19] and enterotoxin [20] are two of the agents so used. After incubation of a liver-cell suspension with pronase, at a concentration of 0.1–0.25% (w/v) for 30–60 min, the parenchymal cells are selectively destroyed and the non-parenchymal cells can then be recovered in proportions equal to those of the intact liver, indicating that cells are not selectively lost by this method. Over 50% of the non-parenchymal cells of the initial suspension can be recovered [21]. Although the action of pronase can alter the cell surface and cause loss of receptor molecules, some of the receptors can be recovered by overnight incubation of the cells [22].

Enterotoxin from *Clostridium perfringens*, at concentrations greater than 1 μg/ml severely damages the hepatocyte membranes, while the non-parenchymal cells remain unaffected [23]. The advantages of using enterotoxin in place of pronase are threefold, enterotoxin acts more selectively on the parenchymal cells, has no enzymic activity, and does not modify the surface of non-parenchymal cells. Enterotoxin also works at a concentration 10× lower than that of pronase. If a liver-cell suspension is incubated for 20–30 min in the presence of enterotoxin at a concentration of 10 μg/ml, the parenchymal cells will become leaky without disintegrating.

The phagocytic ability of the Kupffer cells can be utilized to remove them from a suspension of non-parenchymal cells [26]. Colloidal iron particles can be injected into the rat 30 min before perfusion of the liver is carried out and the cells containing the ingested iron can be removed from the suspension by passing it over a magnetized block.

Most types of mammalian cells, when dead or damaged, tend to pellet through density barriers or gradients of Nycodenz or metrizamide prepared to band cells isopycnically, that is, gradients of up to 28% (w/v) Nycodenz or metrizamide [24, 25]. The enterotoxin-treated cell suspension can therefore be centrifuged on a buffered solution of 28% (w/v) Nycodenz or metrizamide, at 200 g for 5 min, which allows the dead parenchymal cells to pellet and the non-parenchymal cells to be collected at the medium/sample interface. The recovery of non-parenchymal cells after enterotoxin treatment is comparable to that found when pronase is used.

Having recovered the non-parenchymal cells in good yield and representative populations of each type, the next step is to try and subfractionate into their discrete types. As these cells are so similar, both in size and their buoyant densities (rather like the mononuclear cells of blood), it is difficult to separate them on density gradients or barriers. Attempts to purify these cell species on continuous, isotonic Nycodenz gradients have indicated the following buoyant densities: endothelial cells, 1.07 g/ml; Kupffer cells, 1.09 g/ml; stellate cells, 1.09 g/ml; parenchymal cells, 1.11 g/ml [26]. However, there is considerable overlap in the densities which leads to a low recovery of purified cells. On preformed Percoll gradients the buoyant densities of the different species of liver cells have been found to be a little lower than in Nycodenz, but the problem of overlapping bands is still present [27–29]. For a review of the methods used for the purification of liver cells on iodinated media see Berg and Blomhoff [26].

Rat-liver cells, isolated by the perfusion method, have been separated on preformed Percoll gradients [30]. The density range was 1.03–1.10 g/ml, the gradient volume 80 ml and the sample volume 15 ml, containing a cell concentration of 2×10^6/ml. After centrifugation at 800 g for 30 min in a swing-out rotor, the hepatocytes banded at close to 1.08 g/ml and the phagocytes, mainly Kupffer cells, banded at buoyant densities of 1.045–1.06 g/ml.

Other separations on Percoll are listed in the *Percoll Methodology Handbook* obtainable from Pharmacia-LKB.

It is obvious that the purification of the various types of liver cell still leaves much to be desired but, as in so much research work, the researchers have to obtain partially purified material to work on in the absence of a good, reliable alternative. It might be that a systematic investigation of the effects of changes in osmolality, similar to that carried out by Bøyum on blood cells, would produce the necessary method.

8.3.2 *Dissociation of tissues*

It is not always possible to obtain single-cell suspensions by whole-organ perfusion methods, e.g. the samples of solid tumours obtained by biopsy or during an operation. In these cases other methods of preparing the suspensions of single cells have to be employed. Some tissues can be dissociated by just slicing them and agitating them in a culture medium to release the cells. Spleen can be treated in this manner. Other tissues may require harsher methods as different tissues vary in their degree of resistance to disruption. All methods involving enzymic digestion are very likely to damage the cells to some degree and it is therefore advisable to keep the time of exposure to the enzyme action as short as possible, while still being able to obtain representative populations of the cells present in the tissue [1,2].

Solid mammary tumours from mice have been reduced to single-cell suspensions by incubation of the finely sliced tissue with collagenase and the different cell types present separated on discontinuous Nycodenz gradients [31]. The cell suspension was layered on top of a four-step Nycodenz gradient of 5, 10, 15 and 20% (w/v) Nycodenz and centrifuged at 800 g for 20 min. Bands were found at each interface and pelleted material was also present. The cells from each band were tested for their tumour-inducing ability by injecting the cells into isogenic mice. The results indicated that the malignant cells were well separated from the non-malignant by this method [31].

The dissociation methods do have the effect of modifying cell surfaces and causing other damage to the cells. Further damage to already weakened cells can then occur during the washing procedures in which the pelleting of the cells subjects them to compressive forces and resuspending the pellets may add to the stress. If the cells are sedimented onto a cushion of density-gradient medium, the compressive forces will be reduced, resuspension will be more gentle and, if the density of the medium is so adjusted, dead cells will pass through the medium to pellet.

Cells can be dissociated to single-cell suspensions from some tissues without resort to enzymatic digestion; examples include some embryonic tissues, spleen tissue and some cells from testicular tissue. In these cases, the tissue sample can be pulled apart with forceps and shaken loose into a suitable incubation or nutrient medium. Schumacher *et al.* [32] have reported the isolation of Leydig cells from non-enzymatically dissociated testicular tissue from mouse.

The sample was centrifuged on a 0–90% preformed, linear Percoll gradient made up in Earle's BSS and containing 0.07% serum albumin. Centrifugation for 20 min at 800 g separated the cells into four bands, with the Leydig cells concentrated in the third band (from the top) resulting in a 25-fold purification of the Leydig cells, which retained morphological integrity and metabolic activity.

8.4 Isolation of plant protoplasts

The preparation of a single-cell suspension of the higher plant cells presents different problems to those encountered with animal tissues. Plant cells are enclosed within a tough cellulose wall which must be removed in order to release the membrane-bound cell which is then called a protoplast. A single-cell suspension of protoplasts may be likened to a single-cell suspension of mammalian cells in that the cellular content is similar, with the important addition of chloroplasts, and, without the protection of the plant wall, they are equally sensitive to osmotic stress. Isolated plant protoplasts are required for use in genetic studies, such as direct gene transfer, whereby foreign DNA is introduced into the cell and subsequently the cells must be able to divide to form colonies. It is clear that, as in the case of mammalian cells, best results will be obtained when the isolation procedure is carried out in a manner causing the least stress to the protoplasts, allowing them to be recovered in good functional condition.

Protoplasts are also used in order to release their cellular contents into a suitable medium prior to the purification of subcellular organelles, without subjecting them to the severe stresses involved in the harsher method of homogenization in a high-speed blender. While the blender can be used to prepare larger amounts of homogenate, the resulting preparation contains a high proportion of broken organelles with subsequent loss of function.

The protocols for the preparation of protoplasts for culture and direct gene transfer [33] and for the isolation of chloroplasts [34] have been reviewed in the literature. A brief description of one method is given here, followed by suggestions for modification of the centrifugation steps which may reduce stress and enhance recovery of the protoplasts.

Finely sliced leaves are soaked for several hours in the presence of the chosen enzyme and the protoplasts released from within the cellulose walls can be shaken into the medium by gentle agitation. After filtering through a 500 μm, then a 200 μm mesh, to remove the large pieces of tissue remaining, the filtrate will contain the intact protoplasts together with cell debris and the contents of broken cells. The separation of the protoplasts from the organelles and debris entails pelleting the cells at 100 g for 5 min and gently resuspending the pellet in a 0.5 M sucrose solution which is overlaid first with a 0.4 M sucrose then a buffer solution. Centrifugation of this discontinuous gradient for 5 min at 250 g allows the protoplasts, which have a low buoyant density, to float up to the buffer/0.4 M sucrose interface, from whence they are harvested using a Pasteur pipette; the cell debris and denser organelles remaining either in a pellet or in a denser region of the gradient.

Protoplasts are easily broken during the isolation procedure and it has been found difficult to obtain them in sufficient numbers to provide high yields of chloroplasts and other subcellular organelles. The earlier discussion of the isolation of mammalian cells from tissues by enzymic digestion, suggests possible modifications of the above protocol to improve the yield of protoplasts. Firstly, the damage caused by pelleting the protoplasts from the incubation medium can be minimized by sedimenting them onto a cushion of medium, such as the 0.5 M sucrose solution, thus reducing the compressive forces of the pellet against the solid centrifuge tube. Removing the supernatant down as far as the cell band would then allow the protoplasts to be suspended in the sucrose solution by agitation which is more gentle than resuspending from a pellet. The second factor is the osmolality of the sucrose solution, 0.5 M sucrose has an osmolality of about 600 mOsm and thus is hyper-osmotic as compared to the normal environment of the plant. The protoplasts will lose water and increase in buoyant density under these conditions, which is a probable cause of appreciable numbers of protoplasts remaining in the pellet or denser areas of the gradient after centrifugation.

These factors indicate that it may well be worthwhile looking to the properties of iso-osmotic solutions to maximize the recovery of proto-plasts, and the non-ionic iodinated medium could provide such improvement. It has been shown that most mammalian cells band, in iso-osmotic conditions, at buoyant densities ranging from about 1.05 to 1.11 g/ml and it is most probable that protoplasts fall within this range. If the osmolality of the particular plant cell is determined, Nycodenz, metrizamide or Percoll solutions can be prepared which are iso-osmotic with the protoplasts, including the necessary salts and buffers. The protocol described below follows closely the methods of Leegood and Malkin [34] and is suggested as the basis for further experimentation.

A solution of 30% (w/v) Nycodenz, including calcium and an MES buffer will have an osmolality of close to 320 mOsm. If the plant tissue is of greater osmolality, the solution can be adjusted by addition of sodium chloride — 0.1% NaCl will add about 35 mOsm to the Nycodenz solution. The filtrate from the digestion should be distributed to an appropriate number of 50 ml centrifuge tubes and the Nycodenz solution introduced under the filtrate using a syringe fitted with a metal cannula, about 2 ml of Nycodenz per tube. The cells are sedimented onto this cushion at 100 g for 5 min. If the supernatants are carefully drawn off to the top of the cell band, the cells can be suspended in the Nycodenz solution by gentle agitation. This should give a final Nycodenz concentration of about 25% (w/v). If aliquots of the stock Nycodenz solution are diluted with buffer solution containing sufficient sucrose, glucose or other media, to provide an osmotic balance (see Section 5.4), Nycodenz concentrations of 5, 10, 15 and 20% (w/v) can be prepared. A discontinuous gradient can be formed with these dilutions and the Nycodenz solution containing the sample underlayered.

Experimentation with centrifugation times and forces, say 5–15 min and

250–1000 g, will indicate the correct conditions for the optimum recovery of the protoplasts. The number of Nycodenz layers can, of course, be reduced and modified in the light of experimental results.

References

1. Freshney, R.I. (1983) *Culture of Animal Cells: Manual of Basic Technique.* Alan Liss Inc., New York.

2. Freshney, R.I. (1986) *Animal Cell Culture: A Practical Approach.* IRL Press, Oxford.

3. Bøyum, A. (1968) *Scand. J. Clin. Lab. Invest.*, **21** Suppl 97, 77.

4. Bøyum, A. (1983) *Scand. J. Immunol.*, **17**, 429.

5. Nelson, R.D., Quie, P.G. and Simmons, R.L. (1975) *J. Immunol.*, **115**, 1650.

6. Roth, J.A. and Kaeberle, M.L. (1981) *J. Immunol. Methods*, **45**, 153.

7. Jarstrand, C., Lahnborg, G. and Berghem, L. (1979) *Acta Chir. Scand.*, **489**, 279.

8. Hill, R.S., Norris-Jones, R., Still, B. and Brooks, D.E. (1980) *Am. J. Haematol.*, **21**, 249.

9. Wakefield, J.S.J., Gale, J.S., Berridge, M.V., Jordan, T.W. and Ford, H.C. (1982) *Biochem. J.*, **202**, 795.

10. Ford, T., Needle, R. and Rickwood, D. (1987) *Blut*, **54**, 337.

11. Ford, T., Graham, J. and Rickwood, D. (1991) *Clin. Chim. Acta*, in press.

12. Gutierrez, C., Bernabe, R.R., Vega, J. and Kreissler, M. (1979) *J. Immunol. Methods*, **29**, 57.

13. Saksela, E., Timonen, T. and Ranki, A. (1979) *Immunological Rev.*, **44**, 71.

14. Ulmer, A.J. and Flad, H.D. (1979) *J. Immunol. Methods*, **30**, 1.

15. Spooner, J., Percy, R.A. and Rumley, A.G. (1979) *Clin. Biochem.*, **12**, 289.

16. Rennie, C., Thompson, S., Parker, A. and Maddy, A. (1979) *Clin. Chim. Acta*, **98**, 119.

17. Berg, T. and Blomhoff, R. (1983) in *Iodinated Density Gradient Media: A Practical Approach* (D. Rickwood, ed.). IRL Press, Oxford, p.173.

18. Munthe-Kaas, A.C., Berg, T. and Seljelid, R. (1976) *Exp. Cell Res.*, **99**, 146.

19. Mills, D.M. and Zucker-Franklin, D. (1969) *Am. J. Pathol.*, **54**, 147.

20. Berg, T., Tolleshaug, H., Ose, T. and Skjelkvale, R. (1979) *Kupffer Cell Bull.*, **2**, 21.

21. Berg, T. and Boman, D. (1973) *Biochim. Biophys. Acta*, **321**, 585.

22. Munthe-Kaas, A.C., Berg, T., Seglen, P.O. and Seljelid, R. (1975) *J. Exp. Med.*, **141**, 1.

23. Tolleshaug, H., Skjelkvale, R. and Berg, T. (1982) *Infect. Immunity*, **37**, 486.

24. Seglen, P.O. (1976) in *Biological Separations in Iodinated Density Gradient Media* (D. Rickwood, ed.). IRL Press, Oxford, p. 107.

25. Berg, T. and Blomhoff, R. (1983) in *Iodinated Density Gradient Media: A Practical Approach* (D. Rickwood, ed.). IRL Press, Oxford, p. 148.

26. Berg, T. and Blomhoff, R. (1983) in *Iodinated Density Gradient Media: A Practical Approach* (D. Rickwood, ed.). IRL Press, Oxford, p. 151.

27. Voss, B., Allam, S., Rauterberg, J., Ullrich, K., Giesselman, V. and Figura, K. (1979) *Biochem. Biophys. Res. Commun.*, **90**, 1348.

28. Page, D.T. and Garvey, J.S. (1979) *J. Immunol. Methods*, **27**, 159.

29. Seglen, P.O. (1979) in *Cell Populations, Methodological Survey (B): Biochemistry* (E. Reid, ed.). Ellis Horwood, UK, Vol. 9, p. 25.

30. Pertoft, H., Rubin, K., Kjellen, L., Laurent, T.C. and Klingeborn, B. (1977) *Exp. Cell Res.*, **110**, 449.

31. Ford, T., Lai, T. and Symes, M.O. (1987) *Br. J. Exp. Path.*, **68**, 453.

32. Schumacher, M., Schaffer, G. and Holstein, A.F. (1978) *FEBS Lett.*, **91**, 333.

33. Shillito, R.D. and Saul, M.W. (1988) in *Plant Molecular Biology: A Practical Approach* (C. Shaw, ed.). IRL Press, Oxford, p. 161.

34. Leegood, R.C. and Malkin, R. (1986) in *Photosynthesis – Energy Transduction: A Practical Approach* (M.F. Hipkins and N.R. Baker, eds). IRL Press, Oxford, p. 9.

9 Preparation of Subcellular Organelles

The membrane-bound organelles of eukaryotic cells consist of nuclei, mitochondria, lysosomes, endoplasmic reticulum and Golgi bodies, together with peroxisomes and various transport vesicles. A suspension of these intracellular organelles is usually obtained by homogenizing a cell suspension, tissue or whole organ in some suitable medium, using a Dounce homogenizer or similar apparatus. The make-up of the homogenizing medium and the degree of force used in the homogenization will depend upon the particular organelle to be purified and will dictate the size of membrane fragments. Homogenization methods for various purposes have been fully described elsewhere [1].

Routinely, the isolation of most organelles from mammalian cells has been carried out using differential centrifugation to prepare crude samples of the particles of interest, followed by rate-zonal or isopycnic centrifugation on sucrose gradients to obtain pure fractions. While sucrose is cheap and readily available and sucrose gradients are widely used for these purposes, they do have some important disadvantages. At the concentrations required to band the organelles, sucrose solutions are very hypertonic and, as the organelles move through denser and denser areas of the gradient, they will collapse more and more as water is removed from the membrane-bound compartments. The changes in buoyant density thus induced cause co-banding of particles that, in their natural environment, have different buoyant densities, and therefore such particles cannot be resolved on these gradients. The high viscosity of sucrose solutions severely retards the sedimentation rates of the organelles and this has to be overcome in many cases by high gravitational forces and long centrifugation periods, both potentially damaging to the organelles.

The nonionic, iodinated density-gradient media, metrizamide and Nycodenz, provide solutions of sufficient density to band organelles at relatively low viscosity and under iso-osmotic, or only slightly hyper-osmotic, conditions. For these reasons, the following description of the preparation of subcellular organelles from mouse-liver makes use of the advantages of the Nycodenz solutions in order to reduce the time and centrifugal fields required to purify each fraction. A further advantage is that the traditional sucrose methods require the use of an ultracentrifuge, whereas the methods described here can all be carried out using a high-speed centrifuge and rotor, capable of producing only up to 75 000 g [2].

9.1 Preparation of nuclei from mouse liver

Nuclei are the largest of the subcellular organelles and a partially purified fraction can be prepared by differential centrifugation in which a crude nuclear pellet is resuspended and recentrifuged a number of times to remove contaminating material. The several washes of the pellet cause increasing loss and damage to the nuclei and the resulting fraction is never wholly pure, although it may well be of sufficient purity for certain purposes.

The following procedure describes the rapid preparation of highly purified nuclei in high yield [37].

A stock solution of 60% (w/v) Nycodenz, containing 1 mM $CaCl_2$, 1 mM $MgSO_4$, 10 mM Tris/HCl (pH7.6), is prepared; the presence of calcium and magnesium is necessary to preserve the integrity of the nuclei. A buffer solution of 10 mM Tris/HCl (pH 7.6), 1 mM $CaCl_2$, 1 mM $MgSO_4$ is also prepared and an homogenizing medium of 0.25 M sucrose is made up in that buffer. Aliquots of the Nycodenz stock solution are diluted with the buffer solution to prepare 40% and 50% (w/v) final Nycodenz concentrations.

All the following procedures must be carried out ice-cold.

Liver from a freshly killed mouse is cut into several pieces and swirled around in ice-cold homogenizing medium to remove as much blood as possible from the tissue before transferring the pieces to 2–3 ml of fresh medium and slicing finely. Sufficient homogenizing medium is added to make up about 5 ml/g of liver and transferred to the homogenizing vessel. Homogenize by 7–8 strokes of the pestle, rotating at about 700 r.p.m. and filter the homogenate through four layers of muslin, into a 50 ml centrifuge tube. After centrifugation in a pre-cooled fixed-angle rotor for 10 min at 1000 g at 4°C, a pellet containing most of the nuclei is found. The nuclear pellet is resuspended by adding 1–2 ml of the buffer solution and checking the final volume of the suspension of nuclei. Adding 1.45 ml of the stock Nycodenz solution to each 1 ml of the resuspended pellet, gives a final Nycodenz concentration of 35% (w/v).

A discontinuous gradient is prepared by underlayering 3 ml of 40% Nycodenz with 3 ml of 50% in a 15 ml centrifuge tube. Overlayer with the nuclear sample and centrifuge in a pre-cooled swing-out rotor at 10 000 g for 1 h at 4°C. During centrifugation, less dense material will float up to the top of the sample layer, the lysosomes and mitochondria, which contaminate the crude nuclear pellet, will sediment to the sample/40% interface while the larger and denser nuclei will sediment to the 40/50% interface. In the relatively short centrifugation time and low gravitational field, macromolecules and microsomes present in the sample layer, will not have time to migrate to the 40/50% interface, even if they have a higher buoyant density than the nuclei, and this allows a population of pure, intact nuclei to be harvested from that interface.

9.2 Separation of mitochondria, lysosomes and Golgi membranes

Homogenates prepared for the purification of mitochondria, lysosomes or Golgi are homogenized in a medium composed of 0.25 M sucrose, 10 mM Tris/HCl, (pH 7.6) and 1 mM EDTA. The Nycodenz gradients used also contain the same concentrations of Tris/EDTA.

The homogenization is carried out as described for the preparation of nuclei but, in this case, the supernatant from the first nuclear pellet is saved and the pellet discarded. The supernatant is centrifuged at 15 000 g for 10 min, the supernatant discarded and the pellet resuspended in about 4 ml of the homogenizing medium.

A 20–40% (w/v) Nycodenz gradient is prepared by the diffusion method, density range 1.1–1.21 g/ml. The mitochondria sample suspension is top-loaded, 2 ml/8 ml gradient and centrifuged for 90 min at 52 000 g_{av}, in a high-speed centrifuge using a swing-out rotor. The gradients should be unloaded by upward displacement and the fractions from each gradient assayed for the presence of mitochondria (marker enzyme; succinate-INT-reductase) and lysosomes (marker enzyme: ß-galactosidase).

The advantage of using a Nycodenz gradient for this separation can be seen by comparing the results obtained if the sample is centrifuged on a 1–2 M sucrose gradient under identical centrifugation conditions.

Figure 9.1 shows the distribution of the mitochondria and lysosomes in each gradient, clearly demonstrating the differences in the degree of resolution of the particles on the two gradients. In the sucrose gradient the lysosomes have co-banded with the mitochondria, while on the Nycodenz gradient they are almost completely separated. *Figure 9.1* shows the distribution in relation to the gradient fractions of each gradient and the mitochondria appear to be banded in equivalent fractions. However, the density profiles of the gradient are quite different, e.g. the mitochondria are banded at a density of 1.15 g/ml on the Nycodenz gradient, but at a density of 1.2 g/ml on the sucrose gradient. This reflects the difference in the buoyant densities due to the osmotic differences between the two media. In the Nycodenz gradient at the banding density of 1.15 g/ml, the osmolality of the medium is near iso-osmotic, while in the sucrose gradient the osmolality is in excess of 2500 mOsm.

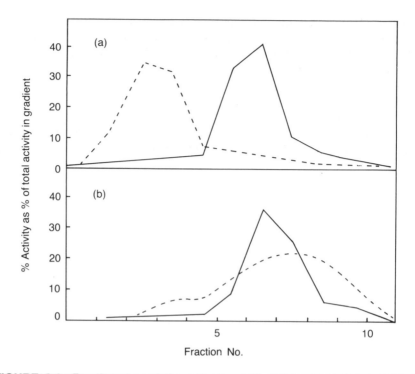

FIGURE 9.1: *Fractionation of the mitochondrial pellet on continuous density gradients. The resuspended mitochondrial pellet was centrifuged on top-loaded (a) Nycodenz 20–40% (w/v) or (b) sucrose 1–2 M gradients for 1.5 h at 52 000 g$_{av}$. Each gradient was fractionated by upward displacement and the fractions assayed for mitochondrial (———) and lysosomal activity (– – –).*

9.2.1 *Effect of changing the conditions of centrifugation*
The resolution of the different particles can be altered by changing the centrifugation conditions, the time and/or r.c.fs used, or the density profile of the gradient. These effects can easily be demonstrated in the following manner.

The crude mitochondrial pellet is resuspended in 60% (w/v) Nycodenz so as to have final Nycodenz concentrations of 25% (w/v) or 40% (w/v). This is best done by resuspending the pellet in 1–2 ml of homogenizing medium then mixing volumes of the suspension and stock Nycodenz to the required concentrations.

The following discontinuous gradients are prepared:

(a) 2 ml 40% Nycodenz, 5 ml sample in 25% Nycodenz, 3 ml 23% and 2 ml each of 20%, 15% and 10% Nycodenz.

(b) 3 ml sample in 40% Nycodenz, overlayered by 2 ml each of 30%, 26%, 24% and 19% Nycodenz. All Nycodenz concentrations are given as % (w/v).

Both gradients are centrifuged in swing-out rotors at 52 000 g_{av} gradient: (a) for 1.5 h and (b) for 2 h. *Figure 9.2a* shows that on (a) an effective separation of mitochondria, lysosomes and Golgi is obtained, while *Figure 9.2b* shows that on (b) the resolution of mitochondria from lysosomes is improved, but there is a reduction in the resolution of lysosomes from Golgi.

It should be clear from the differences in the banding patterns which occur under changes of gradient profiles and/or centrifugation conditions, that the exact protocol required depends upon the characteristics of the particles to be purified. In the procedures described above, the most pure mitochondria fraction appears on gradient (b), while lysosomes and Golgi are better resolved on gradient (a).

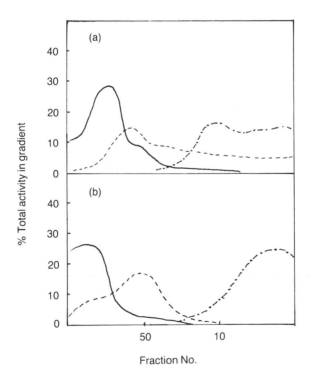

FIGURE 9.2: *The effect of changing centrifugation conditions on the resolution of particles. Two discontinuous Nycodenz gradients were prepared as described in the text. In gradient **(a)** the sample was loaded in the central portion of the gradient, while in **(b)** it was bottom-loaded. Both gradients were centrifuged in swing-out rotors at 52 000 g_{av} (a) for 1.5 h and (b) for 2 h. After fractionation the fractions were assayed for Golgi membranes (———), lysosomes (– – –) and mitochondria (–·–·–).*

It has been explained that, given sufficient centrifugation time and r.c.fs, particles centrifuged on a suitable density gradient will reach a position of equilibrium with the gradient medium and migration will stop. In the case of discontinuous gradients, the particles reach an interface too dense for them to penetrate. Until such equilibrium is reached, the particles are moving at different rates within the gradient and in some cases, in different directions, therefore, during centrifugation particles may pass through stages in which a particular species is in a very pure form in one area of the gradient, but is later contaminated by the migration of other species into that area. Thus, when there are difficulties in separating particles, it may be possible to devise conditions to optimize purity from a knowledge of the sedimentation characteristics of the other particles.

9.3 Isolation of chloroplasts

Chloroplasts, the organelles which carry out the processes of photosynthesis, occur in all photosynthetic organisms except bacteria and the blue–green algae. They are comparatively large organelles with diameters of 3–10 μm, surrounded by a double membrane which contains a gel- like stroma. Within the stroma a system of membranes form a series of fluid-filled sacs stacked in a lamellar arrangement, the thylakoids, which are the site of the light reactions of photosynthesis. The chlorophyll pigments are bound to thylakoid membrane proteins. The stroma of chloroplasts also contains DNA fibrils, (coding for some, but not all, of the chloroplast's proteins) ribosomes and the enzymes of the dark reaction of photosynthesis.

Only the green cells of plants contain chloroplasts and they are most numerous in the palisade and mesophyll tissues, never in the meristematic tissues. The mesophyll is therefore a common starting material for the preparation of chloroplasts, although many workers just use whole leaves.

The manner in which the chloroplasts are isolated depends very much on the subsequent studies to be undertaken. Most studies require intact chloroplasts in the best functional condition possible and for this, very gentle isolation procedures are needed, since they are extremely delicate and easily broken, losing many of their *in vivo* functions rapidly after isolation from the cell. Other studies require the chloroplasts to be isolated in an intact condition in order to examine the structure and function of particular fractions, such as the DNA or ribosomes, free of the influence of other cellular structures. In such cases, loss of certain functions is not important.

The yield and functional condition of a chloroplast preparation depends (apart from the method of preparation) upon the species of plant and the growth regime it has been subjected to prior to harvesting [3]. In this Section, the methods used for these isolation procedures can only be discussed briefly. We have already looked at methods of preparing protoplasts as a source of subcellular plant organelles (Section 8.4). The other methods involve homogenizing the plant tissue in a high-speed blender to release the cellular contents. Here we examine methods of purifying the chloroplasts from tissue broken in a high-speed blender.

It is essential to dark-adapt plants for 12–18 h prior to mechanical disruption of the tissues. This is to disperse starch granules which are present in the chloroplasts and which, during homogenization and centrifugation, can disrupt the chloroplast membranes. If the isolated chloroplasts are to be used for photosynthetic studies, they must then be light-adapted for 30 min prior to harvesting in order to reactivate components of the photosynthetic apparatus.

The harvested leaves are ground in a semi-frozen grinding medium (0.35 M sucrose, 25 mM Hepes, 2 mM EDTA, pH 7.6), about 20 g leaves to 100 ml medium. Two 3-sec bursts at three-quarters full speed grinding should be sufficient. Debris is removed by straining the homogenate through eight layers of muslin. All operations must be carried out as rapidly as possible and at ice-cold temperatures. It is therefore necessary to use a refrigerated centrifuge and to pre-cool the centrifuge and the rotor to be used. The strained homogenate is centrifuged at 4000 g for 1 min to pellet the chloroplasts. The supernatant is discarded and the pellet washed in 10–20 ml of grinding medium and resuspended, using a soft paintbrush. The suspensions of washed chloroplasts so prepared are reported to produce 40–50% intact chloroplasts [4], but it must be remembered that the suspension will also contain nuclei, mitochondria and starch granules as well as lesser amounts of other subcellular fractions. However, this rapid isolation procedure provides large numbers of chloroplasts in sufficiently functional condition for certain subsequent work and, as discussed earlier, there is no point in carrying out prolonged purification operations if partially purified material will suffice. Intact isolated chloroplasts maintain photosynthetic activity, if kept on ice, far longer than thylakoids.

In order to obtain further purification and to separate intact and broken chloroplasts, density-gradient centrifugation is required. Percoll gradients have been designed for this purpose and one is described here.

Aliquots of 10%, 25%, 45%, 65% and 80% (v/v) Percoll in 50 mM Hepes–KOH, 0.33 M sorbitol are prepared; the pH is varied with the requirements of subsequent studies, pH 7.5 for envelope membrane preparations and pH 8.4 if protein transport studies are to be undertaken, for example. The washed chloroplast pellet is resuspended in the Hepes–sorbitol solution, laid on top of the discontinuous Percoll gradient and centrifuged at 1500 g for 15 min at 0°C. The rotor should be allowed to run down without braking. Three chloroplast bands will be apparent, reading from the top, they will consist of (1) broken chloroplasts, (2) intact chloroplasts and (3) aggregated chloroplasts. By this method, chloroplast suspensions 90–95% intact can be obtained.

Methods for the sub-fractionation of chloroplasts for the preparation of stromal extracts, thylakoid membranes and envelope membranes are described by Leegood and Malkin [5].

In this short, introductory book, it is not possible to give further details of the isolation of plant organelles. The isolation of plant nuclei [6] and mitochondria [7] have been described elsewhere. The reader is referred to these texts for more information.

References

1. Graham, J. (1975) in *New Techniques in Biophysics and Cell Biology* (R. Payne and B.J. Smith, eds). Wiley, New York, Vol. 2, p. 1.

2. Graham, J.M., Ford, T.C. and Rickwood, D. (1990) *Anal. Biochem.*, **187**, 318.

3. Morgenthaler, J.J., Marsden, M.P.F. and Price, C.A. (1975) *Arch Biochem. Biophys.*, **168**, 289.

4. Shillito, R.D. and Saul, M.W. (1988) in *Plant Molecular Biology: A Practical Approach* (C. Shaw, ed.). IRL Press, Oxford, p. 161.

5. Leegood, C. and Malkin, R. (1986) in *Photosynthesis – Energy Transduction: A Practical Approach* (M.F. Hipkins and N.R. Baker, eds). IRL Press, Oxford, p. 16.

6. Jofuko, K.D. and Goldberg, R.B. (1988) in *Plant Molecular Biology: A Practical Approach* (C.H. Shaw, ed.). IRL Press, Oxford, p. 37.

7. Schuster, W., Hiesel, R., Wissinger, B., Schobel, W. and Brennicke, A. (1988) in *Plant Molecular Biology: A Practical Approach,* (C.H. Shaw, ed.). IRL Press, Oxford, p. 79.

10 Separation of Macromolecules and Macromolecular Complexes

The relatively small sizes of macromolecules and macromolecular complexes means that their sedimentation rates in density gradients will be very slow in comparison to those of cells and organelles and therefore much longer centrifugation times and higher r.c.fs will be required for them to reach their isopycnic positions on the gradients. Preformed gradients are not necessary in these cases as the time taken for the particles to reach their isopycnic position is greater than the time needed for a self-forming gradient to be generated. It is usual to mix the sample material with the gradient medium to the correct final density, place the mixture into a centrifuge tube and centrifuge for the required time and at the speed needed for the gradient to form and the particles to band.

Sucrose solutions are not used for providing self-generating gradients since the diffusion rate of sucrose is such that it opposes the sedimenting forces of the medium to the extent that a usable gradient is not formed. The colloidal silica media form gradients very quickly, but at the centrifugation forces and times required to band macromolecules, the silica particles will pellet. Colloidal silicas are therefore not used for these purposes. Self-generating gradients of the salts of alkali metals and the iodinated media are commonly used for the separation of macromolecules and macromolecular complexes and their use will be described in the following sections.

10.1 Characteristics of gradients of caesium salts

Caesium chloride and caesium sulfate are probably the most widely used of the salts of alkali metals for density-gradient separations of macromolecules. The dense salt solutions needed do have certain disadvantages however. At the concentrations required, the solutions have very high osmolalities and are of high ionic strength and, while the osmotic stress would only seriously affect osmotically sensitive particles such as cells and organelles, the high ionic strengths of these solutions disrupt most nucleoprotein complexes to nucleic acids and proteins. Thus these solutions cannot be used with complexes such as ribosomes or chromatin unless they are fixed in HCHO to prevent dissociation. Unfortunately, the effect of fixing is irreversible and so the molecules thus treated may be unsuitable for certain investigations to be undertaken. The high concentration of the salt solutions also result in solutions of low water activity, i.e. there are less free water molecules available and thus the macromolecules, especially nucleic acids, become partially dehydrated, the degree of dehydration being related to the water activity of the solution used. This, in turn, means that the macromolecules have a high buoyant density. The density gradient is formed by the increasing salt concentration along the length of the gradient and therefore no control can be exerted over the ionic conditions that the molecules are subjected to within the gradient. The water activity is higher in caesium sulfate gradients than in caesium chloride and thus the buoyant densities of nucleic acids are lower in caesium sulfate solutions; mouse DNA bands at 1.69 g/ml in caesium chloride but 1.42 g/ml in caesium sulfate, while RNA bands at >1.9 g/ml in caesium chloride but at 1.64 g/ml in caesium sulfate. However, on caesium sulfate gradients, RNA tends to aggregate or precipitate, and the gradients formed tend to be steeper and the resolution poorer.

10.2 Characteristics of gradients of the nonionic, iodinated media

The nonionic, iodinated gradient media, metrizamide and Nycodenz, provide solutions of high water activity, as compared with the solutions of caesium salts, and therefore macromolecules can remain almost fully hydrated, thus displaying much lower buoyant densities than those found in the salt gradients. The nonionic nature of their solutions allows the ionic environment of gradients to be varied in a controlled manner by the addition of cations or anions to the gradient solutions. One consequence of this is that the effect of different ions upon the conformation of the macromolecules can be studied under controlled conditions. While the viscosities of solutions of the iodinated media are significantly greater than those of the caesium salts, the lower concentrations of metrizamide and Nycodenz required to band macromolecules (due to their higher degree of hydration) means that the differences in the viscosities are less significant in practice.

Nucleoprotein complexes are not disrupted in the presence of Nycodenz or metrizamide solutions and they can therefore be banded without prior fixing. Approximate banding densities of macromolecules found in gradients of the nonionic, iodinated gradient media are: DNA, 1.12 g/ml; RNA, 1.17–1.2 g/ml; protein, 1.26–1.3 g/ml; carbohydrates, 1.2 g/ml; and nucleoproteins, 1.2–1.35 g/ml.

The choice of media for the fractionation of macromolecules will therefore depend upon the ultimate result required, and while it is not possible to provide a very wide range of protocols in this book, the following sections try to provide sufficient information to allow the selection or design of a suitable technique for particular molecular species.

10.3 Purification of DNA

The banding density of DNA can vary and the variation depends upon a number of different factors, such as the differences between the ratios of base pairing; native DNA and denatured DNA; linear and supercoiled DNA; and, also, differences in the degree of hydration of the DNA. These differences can be exploited to achieve a separation of the various types by the choice of gradient media and the centrifugation conditions employed.

Both native and denatured DNA have slightly higher banding densities in Nycodenz than in metrizamide, probably due to a small decrease in the hydration of DNA in Nycodenz [1]. Native *E.coli* DNA bands at 1.12 g/ml in metrizamide and 1.33 g/ml in Nycodenz, while the denatured form bands at 1.15 and 1.169 g/ml respectively. In caesium chloride gradients, the difference is not so great — native DNA bands at 1.7 g/ml and denatured at 1.715 g/ml — therefore gradient separation is not practical. The separation in caesium sulfate gradients is better and this can be enhanced by the presence of Ag^+ or Hg^{2+} ions in the gradient. These ions bind denatured DNA much more avidly than native DNA, allowing a good separation of the species [2]. *E.coli* DNA, density-labelled with 5-bromodeoxyuridine, shows a significant increase in buoyant density on Nycodenz gradients, banding at 1.175 g/ml [1].

Changes in the buoyant density of DNA due to the binding of ligands depends upon the ligand used and the way in which such binding alters the conformation or degree of hydration of the DNA. Some ligands, such as ethidium bromide, bind by intercalating the stacked bases of the helices [3,4]; others, such as netropsin, are non-intercalating and are thought to bind along the minor groove of the helix [5]. The preferential binding of certain ligands to either AT-rich or GC-rich regions of DNA offers another method of separating DNA species. Differences in the degree of interaction of the intercalating ligands with DNAs of different conformation, e.g. linear and supercoiled molecules [4], has enabled the routine separation of plasmids from bacterial host DNA on ethidium bromide/caesium chloride gradients [6]. While the intercalating ligands have no significant effect upon the buoyant density of DNA in Nycodenz gradients, the non-intercalating, such as netropsin and DAPI, have the effect of increasing the buoyant density. The increase is related to the AT:GC ratio of bases in the molecule; the increase in buoyant density of DNA with AT-rich sequences was significantly greater than that of AT-poor DNA when centrifuged on Nycodenz gradients [7].

10.3.1 *Centrifugation conditions for the purification of DNA*

The separation of plasmid (supercoiled) DNA from host bacterial DNA (linear DNA) on caesium chloride/ethidium bromide gradients is described here as it is probably the most routine and widely used technique.

Plasmid DNA is prepared in a host *E.coli* culture which results in a DNA preparation consisting of a mixture of plasmid and host cell DNA. Due to the supercoiled conformation of the plasmid, the intercalating ethidium bromide is less effective in intercalating the stacked bases of the plasmid than the relaxed conformation of the linear DNA. As a result, the decrease in the buoyant density of DNA when the ligand is bound, is greater in the linear DNA than the supercoiled and two distinct, well separated bands are formed.

The DNA is suspended in buffer solution at a concentration which will give a final gradient concentration of 200–300 μg DNA and caesium chloride is added at a concentration of 1 g/ml and dissolved. The refractive index of the solution should be measured as a check for the correct final concentration and density. The mixture is loaded into the appropriate centrifuge tubes, which should be topped up by addition of caesium chloride solution at a concentration of 1 g/ml, and centrifuged in a fixed-angle rotor for 24 h at 150 000 g, or for 36 h at 100 000 g at a temperature of 15°C. After centrifugation, two bands will be present, separated by 3–5 mm, near the centre of the gradient. There may also be a sticky plug of protein at the top of the gradient and this should be removed before harvesting the DNA bands. The position of the bands can be clearly visualized under UV light and harvested using a Pasteur pipette. A pellet of RNA will also be present.

DNA can be banded on metrizamide or Nycodenz gradients by mixing the DNA sample with the gradient solution to a final medium concentration of 27% (w/v). The medium should contain 10 mM Tris-HCl (pH 7.5) and 1 mM EDTA. The gradient is centrifuged at 63 000 g for 44 h at 5°C in a fixed-angle rotor. The gradient formed is sufficient to band all species of DNA, native, denatured, ligand-bound or density-labelled.

10.4 Purification of RNA

Although there are published methods for the purification of RNA using gradients of the caesium salts, it is probably better to use the nonionic, iodinated media for this purpose, especially to look at the effects of ion interactions upon the buoyant density and conformation of the molecule. On Nycodenz gradients, cytoplasmic RNA can be banded in the presence of different sodium chloride concentrations in a controlled manner in order to observe their effects.

The RNA preparation is mixed with a stock Nycodenz solution (made up in 10 mM Tris-HCl, pH 7.5; 1 mM EDTA) to give a final Nycodenz concentration of 35% (w/v) and centrifuged in a fixed-angle rotor at 63 000 g for 42 h at 5°C. After fractionation of the gradient, the position of the RNA can be determined by assaying each gradient fraction by the orcinol assay [8]. This can be carried out directly on the gradient fractions without the necessity of removing the gradient medium from the sample [9]. If the gradients are prepared with the addition of NaCl, from 0 to 100 mM, the banding pattern of the RNA will be seen to change with the variation of salt concentration. At salt concentrations up to 10 mM, a single peak of RNA is found at a buoyant density of 1.184 g/ml, while at 70 mM salt, a single peak at 1.206 g/ml was evident. At salt concentrations between 10 and 70 mM, two peaks were observed at the above buoyant densities, while increasing the salt concentration to 100 mM caused no further increase in the buoyant density [1].

10.5 Purification of proteins

Ultracentrifugation, as a means of fractionation and analysis of protein molecules, has been largely superceded by electrophoresis and gel filtration, but can still be an efficient method for some applications. Here we consider centrifugation techniques for the separation of viral surface glycoproteins; the analysis of immune complexes between antibody and antigen; and the preparation of lipid–protein complexes, such as serum lipoproteins, which relies more on differences in protein : lipid ratios rather than on protein itself.

The density of most proteins in metrizamide is about 1.27 g/ml, while in caesium chloride the figure is about 1.33 g/ml. Since all proteins have very similar densities, they are normally separated on the basis of size (i.e. differences in sedimentation rate rather than density). Polysaccharides have densities of about 1.62 g/ml in metrizamide and 1.28 g/ml in caesium chloride, so in metrizamide it is possible to separate glycosylated from non-glycosylated proteins on the basis of density. Heavily glycosylated molecules such as proteoglycans would be particularly easy to separate from proteins.

10.5.1 *Separation of viral surface glycoproteins*
The two major proteins of Sendai virus surface membrane, the F protein, responsible for cell fusion, and the Hn protein which has haemagglutination and neuraminidase activities, have been separated by rate-zonal centrifugation [10]. The general approach is as follows.

Once the proteins have been solubilized, normally in a nonionic detergent, such as Nonidet P40® or Triton X-100®, they can be loaded onto a suitable gradient also containing the detergent. Since the differences in sedimentation rates between proteins is small, the sample volume should not exceed 5% of the total gradient volume.

A trial gradient might be a 12 ml, 10–25% (w/v) sucrose gradient in phosphate-buffered saline (PBS) containing 1% Triton X-100. The solubilized protein sample, also in PBS and Triton, must been applied to the top of the gradient in a very sharp band (0.5 ml).

Centrifugation at 100 000 g for 25 h in a swing-out rotor will separate the F and Hn glycoproteins from Sendai virus, the F banding at about 15% sucrose and the Hn at about 25%.

Higher resolution can be obtained if a vertical-tube rotor is used and this has the added advantage of reduced centrifugation time, about 4–6 h.

10.5.2 *Immune complexes*
The work of Steensgaard and his colleagues [11,12] established that the association of monoclonal antibodies and antigen can be studied by rate-zonal centrifugation. Their system can detect the number of antigenic determinants and the number of antibody binding sites. The analysis of the results are quite complex and only a brief introduction to the methodology is possible here.

After incubation of the antibody and antigen in PBS, the sample is layered on top of a 5–20% (w/w) sucrose gradient and centrifuged for 6 h at about 100 000 g. Again, the sample volume must be less than 5% of the gradient volume. Analysis of a number of gradients using different antibody : antigen ratios will reveal two or more peaks (free antibody and one or more complexes). From the size and position of these peaks the nature of the complexes can be deduced.

10.5.3 *Serum lipoproteins*
Although serum lipoproteins are routinely analysed by electrophoresis, their large-scale preparation is best achieved by flotation in a centrifugal field. The different classes of serum lipoproteins (chylomicrons; very low-density lipoprotein VLDL; low-density lipoprotein LDL and high-density lipoprotein HDL) can be isolated in individually separate steps, in which the density of the serum is raised incrementally by the dissolution of an increasing amount of sodium bromide in the serum, or in a single operation using a sodium bromide gradient.

Chylomicrons: centrifugation of the whole serum at 30 000 g for 20–30 min will result in the formation of a milky layer at the surface. This will contain most of the chylomicrons (density <0.95 g/ml).

VLDL/LDL/HDL: because the definition of each class is an operational one, i.e. VLDL, 0.95–1.006 g/ml; LDL, 1.006–1.063 g/ml; HDL, 1.063–1.20 g/ml; it is best left to the operator to decide to what density the serum should be raised to allow the flotation of the desired class of lipoprotein. For example, the closer the density of the serum is raised to 1.006 g/ml, the more VLDL will be recovered by flotation but, at the same time, the more comtamination by LDL occurs. We therefore recommend that the operator chooses the appropriate density and that the density of the serum is checked by weighing after the addition of either solid or a saturated solution of sodium bromide. As a guide, 0.32 g NaBr/ml of serum raises its density to about 1.22 g/ml.

Centrifugation at 100 000 g for about 12 h is sufficient for each class of lipoprotein to float to the surface. Centrifugation may be in swinging-bucket or fixed-angle rotors. The lipoproteins can be recovered using a Pasteur pipette, but the best method is to use a tube-slicer to make a cut just below the floating band as the layer tends to be gelatinous and adheres to the tube wall. A tube-slicer allows the tube wall to be washed with buffer to maximize the recovery of the material.

An alternative method, using a continuous density gradient was developed by Hinton et al. [13]. A continuous density gradient of 0–26.3% NaBr containing 1.1% NaCl and 0.01% EDTA pH 7.8, is generated in a swinging-bucket rotor. The density of the serum sample is adjusted to 1.22 g/ml by addition of sodium bromide and introduced to the bottom of the gradient. The ratio of the gradient : sample volumes should be close to 20:1.

After centrifugation at 90 000 g for 2 h, the VLDL is at the top of the gradient, most of the LDL in the middle third of the gradient and the HDL remains in the sample loading area at the bottom.

References

1. Ford, T.C., Rickwood, D. and Graham, J. (1983) *Anal. Biochem.*, **128**, 232.

2. Summers, W.C. and Szybalski, W. (1967) *J. Mol. Biol.*, **26**, 107.

3. Radloff, R., Bauer, W. and Vinograd, J. (1967) *Proc. Natl Acad. Sci. USA*, **57**, 1514.

4. Bauer, W. and Vinograd, J. (1970) *J. Mol. Biol.*, **47**, 419.

5. Guttann, T., Votavova, H. and Pivec, L. (1976) *Nucleic Acids Res.*, **3**, 835.

6. Birnie, G.D. (1978) in *Centrifugal Separations in Molecular and Cell Biology* (G.D. Birnie and D. Rickwood, eds). Butterworth, London, p. 210.

7. Ford, T.C. and Rickwood, D. (1984) *Nucleic Acids Res.*, **12**, 1219.

8. Schneider, W.C. (1957) in *Methods in Enzymology* (S.P. Colowick and N.O. Kaplan, eds). Academic Press, New York, Vol. 3, p. 680.

9. Rickwood, D., Ford, T.C. and Graham, J. (1982) *Anal. Biochem.*, **123**, 23.

10. Graham, J. (1984) in *Centrifugation: A Practical Approach* (D. Rickwood, ed.). IRL Press, Oxford, p. 244.

11. Steengaard, J., Jacobsen, C., Lowe, J., Ling, N.R. and Jefferies, R. (1982) *Immunology*, **46**, 751.

12. Steengaard, J., Jacobsen, C., Lowe, J., Ling, N.R. and Jefferies, R. (1979) *J. Immunol. Methods*, **29**, 173.

13. Hinton, R.H., Al-Tamer, Y., Mallinson, A. and Marks, V. (1974) *Clin. Chim. Acta*, **53**, 355.

Appendix A. Specialized Centrifugation Techniques

Batch-type zonal rotors

These rotors are of a design completely different to the standard fixed-angle, vertical or swing-out rotors; they are used for large-scale separations of biological particles. The separation chamber is not a tube held within a rotor, but a cylindrical space (commonly of 600–1500 ml capacity). *Figure A1* shows a typical rotor which consists of a rotor bowl and lid and a removable septa (or vane) assembly which divides the cylindrical space into sectors. Commonly the number of sectors is four, sometimes six; the septa running from the assembly core to the wall of the rotor. Channels within the core and the vanes allow access to the centre and either the wall or the floor of the rotor bowl.

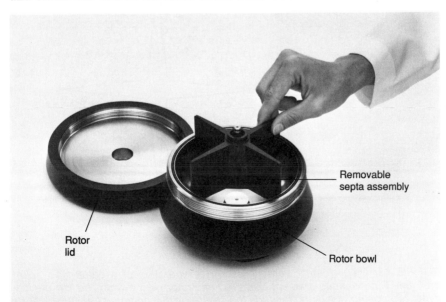

FIGURE A1: Batch-type zonal rotor [courtesy of Beckman Instruments (UK) Ltd].

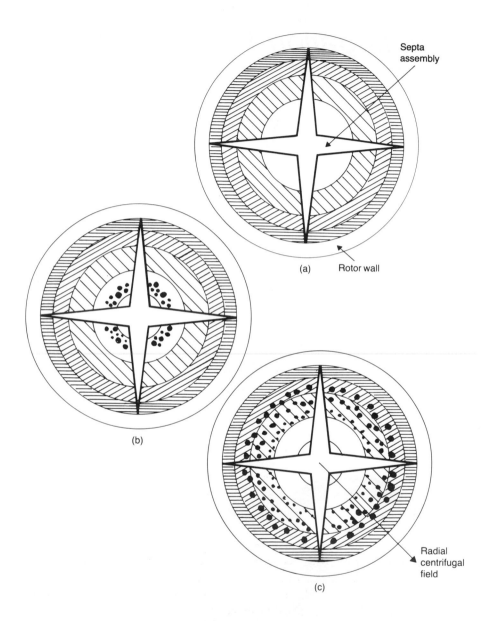

FIGURE A2: *Diagrammatic representation of operation of a batch-type zonal rotor. (**a**) Discontinuous gradient being fed into rotor via the wall. The low density layer ( ) being displaced towards the centre by the incoming denser solutions ( ). (**b**) Sample ( ) fed to the centre of the rotor. (**c**) Particles sedimenting from sample through gradient in an ideal manner.*

FIGURE A3: *Sedimentation of particles through a gradient in a swing-out rotor.*

When the rotor is spinning, a discontinuous gradient appears in the rotor as shown in *Figure A2*, with the least dense layer next to the core and the densest at the wall. Particles in a sample placed adjacent to the core will sediment ideally within each sector-shaped compartment (*Figure A2*). In a tube in a swing-out rotor, only that material in the middle of the tube will sediment along the radial centrifugal field, unimpeded through the gradient (*Figure A3*). Material away from the centre will move firstly towards the wall of the tube before moving further down through the gradient. The problem is less marked, but still present, in a vertical-tube rotor. Only in the sector-shaped compartment of a zonal rotor will all the material in the sample zone sediment ideally.

Zonal rotors are available in the high-speed and ultracentrifuge range. Low-speed models are no longer available commercially. Some models can be loaded and unloaded while the rotor is spinning (dynamically), others must be loaded and unloaded statically. Dynamically loaded and unloaded rotors have a special feed head in place during loading and unloading phases, but not during the separation phase.

The major applications of batch-type zonal rotors reflect their high capacity and potentially high resolution. They have been used widely for the harvesting and purification of viruses from culture fluid. Volumes up to 80% of the total rotor capacity can be processed in a single run. Excellent resolution of ribosomal sub-units and polysomes of different ribosome numbers can be obtained in batch-type zonal rotors and of course they can be used for the scaling up of any tube gradient fractionation: in the zonal rotor the resolution will be at least as good as that in the tube. For more detailed information see Graham [1].

Continuous-flow rotors

These have a design similar to that of batch-type zonal rotors, i.e. they are a hollow cylinder with a central core and septa. In a continuous-flow rotor, however, the core is much wider; the usable space often having a radial distance of less than 2.5 cm. Once the rotor has been loaded with a density gradient, or more routinely, a single or two-part density barrier; a special feed head, which is in place at all phases of the operation, allows a continuous stream of sample to flow up over the core surface and then out of the rotor. As it does so, particles in the stream will sediment out and band close to the wall of the rotor (*Figure A4*). Some models do not permit the use of density barriers (or gradient); in these the particles simply form a pellet at the wall of the rotor.

These rotors are used for the harvesting and partial purification of bacteria and viruses from large (many liters) volumes of culture fluid. For more detailed information see Graham [2].

Elutriator rotors

Elutriation also requires a special feed head to be in place at all times. The kite-shaped sedimentation chamber contains the sample in a suitable medium, through which the particles sediment under the centrifugal field of the spinning rotor. The tendency of the particles to sediment is opposed by a continuous flow of medium which passes in a centripetal direction through the chamber.

During a fractionation, the flow rate of the medium is gradually increased. When the flow rate exceeds the sedimentation rate of the smallest particles, they will be flushed from the chamber in the flowing

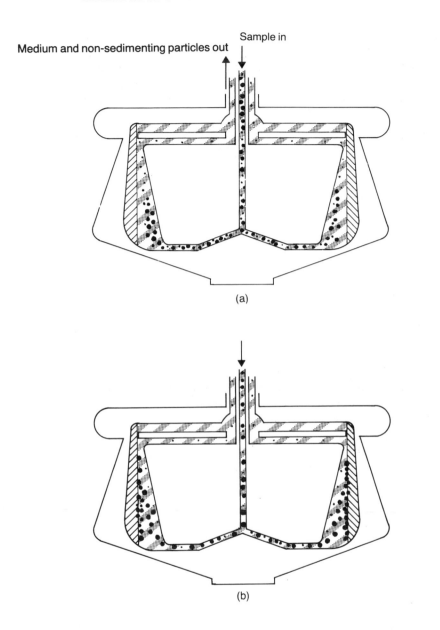

FIGURE A4: *Diagrammatic representation of operation of a continuous-flow zonal rotor. The rotor contains low density medium next to the core (◨) and high density medium next to the wall (▨). The sample feeds to the bottom of the rotor and then flows up over the core surface and out of the rotor. The r.c.f. causes the larger particles (•••) to sediment out of the stream, while smaller ones remain, to emerge at the top of the rotor: (a) early stage of separation; (b) later stage — sample flow maintained and larger particles banding at gradient interface.*

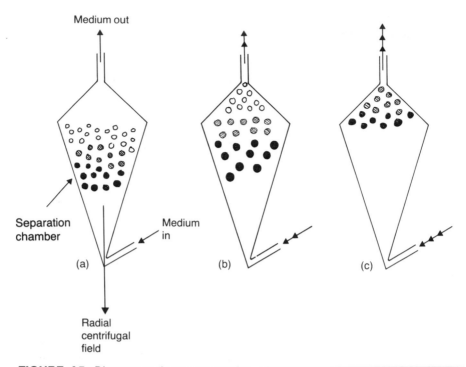

FIGURE A5: *Diagrammatic representation of operation of an elutriation rotor. (a) Chamber contains three different sizes of particles; the outward force of the radial centrifugal field on the particles is balanced by the inward flow of medium. (b) Medium flow is increased, all the particles start to move towards the centre, the smallest ones [oo] exiting from the chamber. (c) The medium flow is increased further to flush the medium size particles [⊚⊚] from the chamber.*

medium, from whence they can be collected. As the flow rate is increased, so progressively larger, more rapidly sedimenting particles will be removed (see *Figure A5*).

The elutriator rotor is accommodated in a high-speed centrifuge and its most popular use is to fractionate cells on the basis of their size.

For further information the reader is referred to more detailed texts in Birnie and Rickwood [3].

Analytical centrifugation

One of the major techniques used for the characterization of protein and nucleic acid molecules in the 1960s and 1970s, was analytical centrifugation. The ellipsoidal-shaped rotor contains up to six holes, close to its circumference, which hold the analytical cells. Each cell is of a complex construction, the sample is essentially contained between two quartz

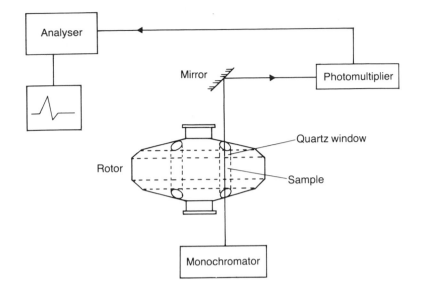

FIGURE A6: *Diagrammatic representation of operation of an analytical centrifuge.*

windows (*Figure A6*) so that, in its simplest form, radiation from a monochromator can pass vertically through the sample, to be detected at a photomultiplier.

The optical system, typically either schlieren or Rayleigh interferometric, detects changes in refractive index within the solution. In this manner the sedimentation of the macromolecule through the medium is monitored and its size deduced. The actual experimental procedures which can involve sedimentation equilibrium analysis in two-component systems are exceedingly complex and far beyond the scope of this text. For more information the reader is referred to Birnie and Rickwood [3].

References

1. Graham, J. (1984) in *Centrifugation: A Practical Approach* (D. Rickwood, ed.). IRL Press, Oxford, p. 226.

2. Graham, J. (1984) in *Centrifugation: A Practical Approach* (D. Rickwood, ed.). IRL Press, Oxford, p. 230.

3. G.D. Birnie and D. Rickwood (1978) *Centrifugal Separations in Molecular and Cell Biology*. Butterworth, London.

Appendix B. Commercially Available Centrifuges

Ultracentrifuges

Manufacturer	Model	Type	Max r.p.m. ×1000	Capacity
Beckman	L7	GF	35,55,65	A
	L8M	HF	60,70,80	A
	Optima L	GF	60,70,80	A
	Optima XL	HF	70,80,90	A
	TL100	HB	100	B
Kontron	T1000	GF	45,55,65,75	A
	T2000	HF	60,70,80	A
Sorvall	OTD	HF	80	A

Type code:
   B Bench-top
   F Floor-standing
   G General purpose
   H High specification

Ultracentrifuges are often available in two models: a general purpose preparative machine with an option of three or four maximum speeds, and a more sophisticated high specification model whose speed options are higher.

Capacity code:
   A Up to approx 600 ml. Only fixed-angle rotors have the highest capacity. Only the smallest capacity fixed-angle rotors are capable of spinning at the highest r.c.f. (up to 600 000 g). Swing-out rotors normally have maximum capacities of about 100 ml: highest r.c.f. (up to 400 000 g) for smallest capacity rotor.

   B Individual samples 0.5–4.0 ml. Unusually the highest capacity fixed angle rotor (21 ml) has the highest r.c.f. (540 000 g).

High-speed centrifuges

Manufacturer	Model	Type	Max r.c.f.[a] ×1000 FA	SO	Capacity
Beckman	J21	FR	52	27	A
Burkard	µP series	BRL	32	15	B
Heraeus	Suprafuge 22	FR	60	27	A
	Varifuge 20RS	FR	48	20	A
	Biofuge 17RS	BRL	27	3	B
Hermle	ZK401	FR	70	26	A
	ZK380	FR	22	4	B
	Z380	B	22	4	B
	Z360K	BR	22	4	B
Howe-Sigma	3K−20	BRL	42		C
	2K−15	BRL	20		C
	3K−12	BRL	12	5	C
	2−15	BL	16	3	C
	4K−10	BRL	18	5	B
IEC	B22(M)	FR	49	28	A
	Centra 4B	BRL	13	3	B
	Centra SR	BRL	17		C
Jouan	MR14.11	BRL	14		C
Kontron	T124/T224/T324	FR	69	26	A
Sorvall	RC5 series	FR	49	45	A
	RC28S	FRX	100	74	B(A)

[a] The maximum r.c.f. is quoted for both fixed-angle (FA) and swing-out (SO) rotors. Only the smallest capacity rotors of either type can achieve the maximum.

Type code:
  B Bench-top
  F Floor-standing
  L Limited range of sample types and sizes. Swing-out rotors either not available or capable only of low speeds
  R Refrigerated or refrigeration optional
  X Rotors for RC5 type centrifuges operate in the RC28S up to 20 000 r.p.m. (i.e. it performs as a capacity A high-speed centrifuge). Other rotors are available (capacity B) which can operate up to 28 000 r.p.m.

Capacity code:
  A 1–3 liters
  B 300–900 ml
  C below 300 ml

Low-speed centrifuges

Manufacturer	Model	Type	Max r.p.m.[a] ×1000	Capacity
Beckman	J6	FRS	6.4	A
	GP Series	BRS	6.4	B
Clandon	T51	BL	7.0	E
	T53	B	10.0	B
Denley	BS400/BR	BRL	6.0	D
Heraeus	Varifuge F	FRS	6.0	B
	Varifuge 3.2S	FRS	6.0	B
	Minifuge F	FR	6.0	C
	Cyrofuge 6000	FRS	4.0	A
	Cryofuge 8500	FRS	5.0	A
	Megafuge	B	6.0	C
	Omnifuge 2.0RS	BRS	6.0	B
	Labofuge	BL	5.3	D
	Medifuge	BL	4.8	E
Hermle	ZK630	FRX	5.2	A
	Z510	BXY	9.2	B
	ZK510	FRXY	9.2	B
	Z320	BY	10.0	E
Howe-Sigma	6–10	BR	9.0	B
IEC	C-6000	FRS	6.0	B
	C-7000	FRS	7.0	A
	Centra 4R	BRS	6.9	B
	Centra 2	B	6.0	D
	Centra 3C	B	6.0	C
	Centra HN	B	4.9	D
	Centra B	BL	3.3	E
	Spinette	BL	3.2	D
Jouan	BB	BL	5.0	E
	B/BR 3.11	BRL	6.6	D
	C/CR 3.12	BR	6.6	C
	C/CR 4.12	BRS	8.0	B
	C/CR 4.22	BRS	8.0	B
MSE	Mistral 6000	FRS	6.0	A
	Mistral 1000	B	6.0	C
	Mistral 2000	B	6.5	C
	Mistral 3000	BRS	6.0	B
Sorvall	RC3	FRS	6.0	A
	Omnispin	FRS	3.6	B

[a] The quoted maximum r.p.m. is usually available only on the smaller fixed-angle rotors, although a few swing-out rotors may be able to achieve this figure. At the highest speed, fixed-angle rotors may generate up to 9000 g, although 5–6000 g is more normal. Swing-out rotors rarely generate more than 5000 g.

Type code:
  B Bench-top
  F Floor-standing
  L Limited range of sample type and size
  R Refrigerated or refrigeration optional
  S Large range of sample types and sizes
  X No fixed-angle rotors available
  Y The maximum quoted speed is only attainable either with microfuge rotors or drum-type (tubes fixed in horizontal position) rotors

Capacity code:
  A up to 6 liters
  B 2–4 liters
  C 600–1500 ml
  D 500–600 ml
  E below 500 ml

Appendix C. Commercially Available Rotors

The rotors for low-speed machines are too diverse to be listed in any useful and digestible form. Indeed even for high-speed and ultracentrifuges it is impractical to list all the rotors available. Instead, for each rotor type (fixed-angle, swing-out and vertical), the r.c.f. and capacity ranges are given for each manufacturer. Alternatively the characteristics of two or three selected rotors which span the capacity range will be given. Generally, only the smallest capacity rotors will be capable of the highest r.c.f. For ultracentrifuge rotors the range of k factors is also given.

The k factor provides an estimate of the time (t) required to pellet a particle of known sedimentation coefficient (s) at the maximum speed of the rotor (see *equations C1* and *C2*).

(C1)
$$t = \frac{k}{s_{20,w}} .$$

For two different rotors 1 and 2:

(C2)
$$t_1 = \frac{k_1 t_2}{k_2} .$$

The k factor is proportional to the ratio of r_{max}/r_{min} for the rotor, i.e. the larger the r_{min} and the smaller the sedimentation path length, the smaller will be k. Clearly from *equation C1*, the smaller the k factor, the more efficient the rotor. *Equation C2* allows the operator to extrapolate the centrifugation time for one rotor to that for another.

In the following tables the capacity is given as tube number × tube volume: many rotors can be adapted to take a range of tube sizes.

Most high-speed machines have rotors which can be adapted to take microtubes or specific microtube rotors. The latter are not listed, neither

Caution: the use of some adapters imposes a restriction on the maximum speed of the rotor.

are centrifuges which accommodate only small sample volume rotors (less than 5 ml tube size).

The quoted maximum speeds are for polyallomer or polycarbonate tubes: it may not be permissible to run some other tube materials at this speed.

Not all rotors may be capable of running in all a manufacturer's machines.

High-speed centrifuges

For sample volumes in excess of 2 ml, only certain fixed-angle rotors in bench-top high-speed centrifuges are capable of high speeds. With the exception of one rotor, all swing-out rotors are restricted to low speeds. Drum rotors, in which the sample is held horizontally may be capable of high speeds, although these are normally used in conjunction with small sample volumes. The restrictions which capacity places on speed are shown in the relevant table. For each manufacturer, three fixed angle rotors have been chosen, their maximum r.p.m. and capacity are given in sequence. Only swing-out rotors accepting tube volumes greater than 2 ml are listed: two examples are given, if available.

Ultracentrifuges

With the advent of very high-speed ultracentrifuges, not all the rotors available can achieve the highest speeds: fixed-angle ultracentrifuge rotors have therefore been divided into three speed categories and swing-out rotors into two. Again individual rotors are not listed, instead the range of capacities and the equivalent range of k factors are given.

High-speed centrifuges (floor-standing)

Manufacturer	Max r.c.f.[a] × 1000	Capacity range (tube number × tube volume)	No. of rotors
Fixed-angle rotors			
Beckman	17–51/44	6×500(72×3.5)–32×15(32×8.0)	7
Heraeus	26–46/44	6×500(54×3.5)–32×16(24×5.0)	7[b]
Hermle	14–71/57	6×500–24×15	7
IEC	14–48/42	6×500(72×5.0)–24×16(24×5.0)	6[b]
Kontron	33–70/57	6×290(90×2.0)–24×16(24×1.0)	7[c]
Sorvall	28–49/50	6×315 (72×3.0)–26×16(24×3.0)	7[bc]
	100	16×36(16×0.4)–16×14(16×0.4)	2[d]
Swing-out rotors			
Beckman	10–27	4×250(48×3.5)–6×50(36×1.5)	2
Heraeus	10–28	4×290(12×3.5)–4×94(12×3.5)	2[b]
Hermle	9–26	4×250(24×5)–4×94(4×50)	2
IEC	10–28	4×250(48×5.0)–4×50(12×5.0)	2[b]
Kontron	9–27	4×250(48×4.0)–4×50(16×1.0)	3[e]
Sorvall	9–45	4×250(48×4.0)–8×10	3[b]
	58–74	6×20(6×0.4)–6×17(6×0.4)	3[d]
Vertical rotors			
Beckman	42	8×39(8×15)	1
Hermle	36–40	8×94–8×38.5	2
Kontron	37–40	8×94(8×4.4)–8×38.5(8×4.4)	2
Sorvall	40–41	8×36(8×17)–6×5	2

[a] A max r.c.f. designation x/y indicates a fixed-angle rotor with two concentric circles of tube pockets, x is the r.c.f. of the outer circle and y that of the inner.

[b] Some of these rotors also have adapters for microfuge tubes.

[c] The largest volume fixed-angle rotor (6 × 500 ml with a max r.c.f. of 14 000 g) cannot be adapted to take smaller tubes.

[d] Rotors for the Sorvall RC28S 'Supra'-speed centrifuge only; this machine will, however, also accommodate all the rotors listed for the Sorvall high-speed centrifuge.

[e] Other large capacity rotors are only capable of low speeds.

Figures in parentheses give the number and volume of the smallest tubes for which a rotor can be adapted. For example, the first line of the table is read as follows: Beckman manufacture seven fixed-angle rotors for their high-speed (floor-standing) centrifuge. The largest capacity rotor is 6 × 500 ml which can be adapted to take a range of tube sizes down to 72 × 3.5 ml: it has a maximum r.c.f. of 17 000 g. The smallest capacity rotor is 32 × 15 ml which can be adapted to take tube sizes down to 32 × 8.0 ml; it has two concentric tube pockets, the outer one has a maximum r.c.f. of 51 000 g, the inner one 44 000 g.

High-speed centrifuges (bench-top)

Manufacturer	Max r.p.m. × 1000	Capacity (ml)	No. of rotors
Burkard	5; 15; 20	6×100; 18×12; 48×0.4	11[f]
	4; 12	4×100; 4×10	2[s]
Heraeus	11; 15; 17;	6×94; 12×13; 40×0.4	3[fa]
	4.3	4×100	1[s]
Hermle	14	6×94; 8×30	2[f]
	4.8; 5.0	4×190; 4×100	2[s]
Howe-Sigma	9; 15; 16	4×250; 6×85; 20×10	7[f]
	4.5; 5.5	4×100; 48×15	4[s]
IEC	5; 14; 15	8×50; 12×10; 24×1.5	13[fa]
	4.3; 5.2	4×175; 20×5	2[s]
Jouan	11; 12; 12	8×38; 10×10; 28×1.5	5[f]

[f] Fixed angle rotors.
[a] Individual rotors (usually the larger capacity ones can be adapted to take smaller volume samples.
[s] Swing-out rotors can usually be adapted to take smaller volume samples.

Ultracentrifuges (floor-standing)

r.p.m × 1000	Manufacturer	Capacity range (tube no. × vol.)	k factor range	No. of rotors
Fixed-angle rotors				
50–90	Beckman	8×13.5–12×38.5	25–69	13[a]
	Kontron	12×2.0–12×38.5	23–51	12
	Sorvall	12×2.0–12×38.5	23–72	9
40–49	Beckman	72×0.2–6×94.0	9–133	4
	Kontron	20×6.5–6×94.0	45–119	2
	Sorvall	40×6.5–6×100	45–162	4
<40	Beckman	100×1.0–6×250	71–951	7
	Kontron	32×13.5–6×250	186–628	2
	Sorvall	6×250	851	1
Swing-out rotors				
40–65	Beckman	3×5.0–6×14.0	54–137	6
	Kontron	6×4.4–6×14.0	45–126	3
	Sorvall	6×4.4–6×13.2	45–123	3
<40	Beckman	6×8.0–6×38.5	138–231	4
	Kontron	6×17.0–6×38.5	233–272	2
	Sorvall	6×17.0–6×36.0	222–273	3
Vertical rotors				
70–90	Beckman	8×5.1	6–8	2
50–65	Beckman	16×5.1–10×39.0	11–36	6
	Kontron	8×6.0–8×35.0	21–36	3
	Sorvall	8×5.0–8×36.0	10–36	3
Nearly vertical rotors				
65–90	Beckman	8×5.1–8×13.5	10–21	2

[a] This group of rotors includes a 44 × 6.5 ml rotor.

Ultracentrifuges (bench-top) (Those rotors are for the Beckman TL100 machine.)

r.p.m × 1000	Rotor type	Capacity range (tube no. × vol.)	k factor range	No. of rotors
100	Fixed-angle	20×0.2–6×3.5	6.6–16.5	3
45	Fixed-angle	12×1.5	99	1
55	Swing-out	4×2.2	50	1
100	Vertical	8×2.0	9.5	1
100	Nearly vertical	8×3.9	14	1

NB Most of the rotors, except for the smallest, can be adapted to take smaller than the quoted volumes.

Appendix D. Further Reading

Having now come to the end of this introductory book on the techniques and applications of centrifugation, we sincerely hope that readers have found the contents interesting and informative. We have tried to present the important aspects of the centrifugation techniques in a manner useful for the day-to-day, routine operation of the centrifuge. At the same time, we have attempted to provide sufficient information to allow the design of new applications without going into detailed descriptions of the mathematical theory of sedimentation. In this way, we have avoided presenting a formidable array of equations, which those new to centrifugation may have found rather daunting and which is unnecessary for most centrifugation work. However, for those wishing to pursue the subject further, the following texts will prove useful.

Birnie, G.D. (1972) *Subcellular Components, Preparation and Fractionation*. 2nd edn. Butterworth, London.

Birnie, G.D. (1976) *Subnuclear Components, Preparation and Fractionation*. Butterworth, London.

Birnie, G.D. and Rickwood, D. (1978) *Centrifugal Separations in Molecular and Cell Biology*. Butterworth, London.

Hinton, R.H. and Dobrota, M. (1976) *Density Gradient Centrifugation*. North-Holland Publishing Co., Amsterdam.

Howard-Evans, W. (1978) *Preparation and Characterisation of Mammalian Plasma Membranes*. North-Holland Publishing Co., Amsterdam.

Rickwood, D. (1976) *Biological Separations in Iodinated Density Gradient Media*. IRL Press, Oxford.

Rickwood, D. (1984) *Centrifugation: A Practical Approach*. 2nd edn. IRL Press, Oxford.

Rickwood, D. (1983) *Iodinated Density Gradient Media*. IRL Press, Oxford.

Booklets, application charts and reference lists may be obtained from the following suppliers.

Nycomed Pharma A/S, PO Box 4284 Torshov, 0401 Oslo 4, Norway.

Pharmacia-LKB, Biotechnology International AB, S-571 82, Uppsala, Sweden.

Appendix E. Major Suppliers of Centrifuge Equipment and Media

Centrifuges

Beckman Instruments Inc.
Spinco Division
1050 Page Mill Road
Palo Alto, California 94304
USA

Clandon Scientific Ltd
Lysons Avenue
Ash Vale
Aldershot
Hants GU12 5QR
UK

Du Pont Company
Sorvall Division
Concord Plaza – Quillen Bldg
Wilmington, Delaware 19898
USA

Berthold Hermle GmbH
Postfach 1240
Industriestrasse 8–12
D-7209 Gosheim
FRG

Jouan SA
Case Postale 3203
Rue Bobby Sands
44805 Saint-Herblain Cedex
France

Burkard Instruments
Biotech Instruments Ltd
183 Camford Way
Luton
Beds LU3 3AN
UK

Denley Instruments Ltd
Natts Lane
Billingshurst
West Sussex RH14 9EY
UK

Heraeus Sepatech GmbH
PO Box 12 20
Am Kalkberg
D-3360 Osterode
FRG

International Equipment Company
(IEC)
300 Second Avenue
Needham Heights
Mass 02194
USA

Kontron Instruments
Via G Fantoli 16/15
20138 Milano
Italy

MSE
Bishop Meadow Road
Loughborough
Leicestershire LE11 0RG
UK

Sigma Laborzentrifugen GmbH
Postfach 1727
D-3360 Osterode am Harz
FRG

Centrifugation accessories

Nalge Company
Box 20365
Rochester
New York 14602-0365
USA

Nycomed AS
Pharma Diagnostica
Sandakervn 64
PO Box 4284 Torshov
N-0401 Oslo 4
Norway

Seton Scientific
PO Box 60548
Sunnyvale
California 94088
USA

Centrifugation media

ICN-Flow
3300 Hyland Avenue
Costa Mesa
California 92626
USA

Nycomed AS
Pharma Diagnostica
Sandakervn 64
PO Box 4284 Torshov
N-0401 Oslo 4
Norway

Pharmacia Fine Chemicals
Box 175
S-751 04 Uppsala 1
Sweden

Sigma Chemical Company
Fancy Road
Poole
Dorset BH17 7TG
UK

All the centrifuge manufacturers supply tubes and accessories. In addition many general scientific equipment companies supply bench-top low-speed and high-speed centrifuges (particularly small capacity machines), accessories and centrifuge tubes.

Products which are registered trade marks of any of these companies are denoted as such at their first mention in the text.

Index